Re/defining the Matrix

Transpekte:
Transdisziplinäre Perspektiven
der Sozial- und Kulturwissenschaften

Transpects:
Transdisciplinary Perspectives
of the Social Sciences and Humanities

Herausgegeben von Johannes Angermüller,
Anke Bartels, Dietmar Fricke, Raj Kollmorgen,
Jörg Meyer, Dirk Wiemann

Bd./Vol. 3

PETER LANG
Frankfurt am Main · Berlin · Bern · Bruxelles · New York · Oxford · Wien

Anke Bartels/Michael Schultze/
Agata Stopińska
(eds.)

Re/defining the Matrix

Reflections on Time – Space – Agency

PETER LANG
Europäischer Verlag der Wissenschaften

Bibliographic Information published by Die Deutsche Bibliothek
Die Deutsche Bibliothek lists this publication in the Deutsche Nationalbibliografie; detailed bibliographic data is available in the internet at <http://dnb.ddb.de>.

Cover illustration:
Anton Giuglio & Arturo Bragaglia: Typist (1911)

ISSN 1860-868X
ISBN 3-631-55585-7
US-ISBN 0-8204-8706-6

www.peterlang.de

Contents

Agata Stopińska, Michael Schultze & Anke Bartels
(Un)Tying the Knot: Time – Space – Agency. An Introduction 7

Alternative Times and Postmodern Mapping

Dagmar Reichardt
Mapping Sicily: From Postcoloniality to Neo-Metaphorization 17

Maria Antonietta Mariani
Engraving Space: Maps, a World-Body Portrait 31

Hogne Øian
Linear Time: The Enemy of the Unemployed 39

Steven Arxer, John Murhpy & Linda Belgrave
Globalization and the Politics of Time 51

Agency and Resistance

Agata Stopińska
Genders in Flux: Polish and Irish Velocities of Doing Gender 63

Orly Benjamin & Sarit Sambol
Surviving Spatial Oppressions Over Time: Israeli Working Poor Women's Localized Resistance 77

Agnieszka Kołodziej-Durnaś
Time, Space and Agency in a Life Project: Polish Students' Reflexivity 89

Catarina Frois
Identification and Anonymity: Two Sides of the Same Coin 103

Steven Corcoran
Postulating Equality: Politics and Philosophy in the Oeuvre of Jacques Rancière 115

Aesthetic Representations

Dirk Wiemann
Mountaineering with Hobbes and Milton: (R)Enunciations of Seeing and Agency in Restoration England 131

Jatin Waglé
Resistant Forms and Critical Subjectivities: Reading Adorno in Our Times and Other Languages 149

Leena Petersen
Poetics of the Intrinsic: On Walter Benjamin's Concept of Allegory 163

Rocío Orsi Portalo
Greek Tragedy and Poetic Agency 177

Ulf Schulenberg
Democratic Space and Poetic Agency: Emerson, Whitman, and the (Im)Possibility of Dissent 187

Agency and Its Spatio-Temporal Dimensions

Holger Rossow
Time-Space Compression, Flexible Accumulation and the Problem of Political Agency in the Age of Globalisation 203

Tarja-Liisa Laaksonen
Situating Liberal "Organizational Development" and "Good Governance" Discourses in Globalisation and Political Contexts 219

Jörg Meyer
Agency Lost: Scientific Respectability and Real World Relevance as Pitfalls of Contemporary International Thought 229

Contributors 245

Agata Stopińska, Michael Schultze & Anke Bartels

(Un)Tying the Knot: Time – Space – Agency

An Introduction

Postmodern interventions have effectively done away with the traditional modern notion of the stable subject as a self-centred, rational and ideally 'sovereign' active agent. But how can agency, which is embedded in concrete spatio-temporal configurations, be conceived after the modern subject? Does the postmodern reformulation of the subject as split, as a site of lack or as an effect of discourse entail disempowerment? Can the sense of agency still be interpreted as an awareness of the ability to initiate, control, and challenge behaviour in terms of the individual's competence merely circumscribed by certain social frameworks in specific spatio-temporal dimensions? Or, can agency – as a concept – be understood as a simultaneity of the options to act and the positions, statuses and roles from which to act?

The main problem in discussing agency seems to lie not only in the difficulty to name and discern points of intervention which might lead to a transformation of social systems but also in identifying agents or collectivities who are capable of carrying out these interventions as well as in the problem of understanding what makes people into agents or groups into collectivities. Moreover, the concept of agency may also carry the burden of distinguishing between the subject and the object, the actor and the spectator, the producer and the recipient – notions that are contingent on the angle of perception or the understanding of the constitution of the social world one agrees to. Lastly, agency is often obscured by a focus on hegemonic resistance which reflects only a part of its meaning embedded in the notion of *intentionality*.

While in some quarters the local capacity to resist or subvert certain aspects of the dominant culture has been celebrated and taken as proof that power relations and existing structures of meaning are not as fixed and stable as formerly implied, there is a danger involved in this kind of reasoning. By emphasising the categories of 'choice' and 'decision', the effectivity and pervasiveness of the existing power structures may be denied and, at the same time, the potential for a mere renewal of existing hegemonic normativity may not be taken into account. This, in turn, prevents a thorough understanding of the workings and effects of power, authority and rule as well as the accompanying or constitutive hegemonic discourses. Also, it might result in a thorough de-politicisation. Thus, neo-liberal capitalism is often used as a metaphor for a complete homogenisation which prevents agency – in the sense of agency as the production of something new, something beyond neo-liberal capitalism – as well as the formation of dissenting alliances which appear to be outside the *only* logic of evolution of human society. Furthermore, neo-liberalism has the tendency to evade certain notions that might point to alternative concepts and discourses by using rather euphemistic ones. In this way neo-liberal discourses contain 'soft methods' of ignoring the possibility of unwanted agency by suppressing the articulation of opposing voices.

To counter this tendency, it seems necessary to relocate the possibilities for agency by an understanding of power that provides the terms for its own critique and thus allows to escape the controlling discourses of the dominant order by means of transgression or opposition. Furthermore, it seems to be necessary to point to a concept of agency, which insists on the capacity to change any form of rules and rule no matter how powerful they are and how natural or rational they appear. At the same time any analysis of agency changing or preserving existing social arrangements should also reflect on the responsibility for the consequences of keeping things as they are or changing them.

Agency does not only open up possibilities to act but is also limited in several respects. Some of these limits are spatial or temporal. Even though any social arrangement, regime or society sets or contents these limits of and to agency, they can be criticised, altered or discussed. This especially holds true for the praxis of neo-liberal globalisation.

While this new, allegedly all-incorporating spatio-temporal regime restricts critical or dissident modes of thinking, as well as deviant and insurgent subjectivities, the possibilities for imagining and creating alternative times, places and ways of life need to be articulated even if this might prove difficult. It can be done by redefining the formation of spatial boundaries, patterns of individual and collective movements, the distribution of power, the creation of sacred and profane, open and closed, private and public spaces in which agency takes place, as well as redefining the meanings of the term 'space'. Today, space seems to become less and less important. With goods now moving from place to place, we no longer make journeys but are on vacation at a certain place, there are no white spots to discover. But still, space makes order. We did and do conceptualise and categorise our world in(to) distinct spaces with boundaries which, again, limit agency. The most obvious visual expression of these conceptualisations and categorisations can be found in maps, which in turn shape our understanding of the world we live in.

We can find these tensions concerning the meaning of the term or the concept of 'space' in various aspects of the social and political life. Inter alia, the concept of sovereignty which was bound to distinct territories and boundaries is now extended beyond these clearly defined spaces. At the same time, the concept of 'time' shifts away from something which is experienced, diverse or flexible, and follows a logic of events (like winter time or time of harvest) into an understanding of 'time' as standardised, homogenised, and 'empty'. Consequently, agency becomes even more limited when it has to be in line with a straight and rationally objectified time because any alternative understanding of 'time' would then be distracting, irrational or subjective.

But at the same time as the concepts of 'space' and 'time' become homogenised they become in a sense fluid. In today's era of unlimited possibilities of travel across the globe within the shortest time ever, together with the accelerating pace of our lives, this fluidity has been changing the concepts of localities and temporalities. These changes are coupled with the interconnectedness of places and the facilitation of agency, and even if this spatial heterogeneity does not create new openings for new meanings immediately, there is a clear flux in the perception of 'time' and 'space' in individual lives. Therefore, thinking through "lifeworlds" today means, for both individuals and whole societies, different reflections on place, one's lifespan, and the attributes of 'time' from those of a century ago. Places, before collectively made meaningful, get a different significance in contemporary cultures.

The contributions collected in this volume reflect the multifaceted discussion of questions concerning spatio-temporal regimes and subjectivities/ agency and form a selection of the papers presented at the international interdisciplinary conference on *Time – Space – Agency*, held at the University of Magdeburg in July, 2005 and sponsored by the *DFG (German Research Foundation)*. The conference aimed at bringing together different approaches and disciplines (e.g. cultural studies, philosophy, political science, literary studies) in order to facilitate the interaction between the social sciences and the humanities. This volume is subdivided into four parts which reflect the main strands of the conference discussions.

The first chapter, *Alternative Times & Postmodern Mapping*, focuses on spatial regimes and cartography as well as dissenting concepts of 'time' and their relevance for present and future possibilities of agency.

Dagmar Reichardt uses the example of Sicilian cartography and iconic articulation to explore theoretical aspects of cultural representation. She illustrates how the oppressive status of postcoloniality and the opposing tendency towards metaphorisation have characterised the image that we associate with Sicily today. The postcolonial concept of a symbolic "mapping of the world" is of specific interest here because of an increasing tendency towards neo-metaphorisation that characterises the academic landscape.

Maria Antonietta Mariani gets deeper into the hidden space-time relation while investigating the drawing scale of maps. In her view, since maps multiply and altern the experience of 'time', they map abstract a space dimension that keeps time and distorts it. Mariani perceives a map as a world-body portrait, which – just like a body – is tied to spatial organisation. She reflects on the destiny of maps, of their construction, shapes and signs, implying that they represent the structure of a society and a vision of the world. Mariani fathoms maps as an instrument for measuring differences and a supporting tool for imagining visible and invisible borders. Mariani studies mapped places that are no longer unknown places, she looks at maps as at an image of political and economical power and examines how the interaction between the objects represented by maps and their readers' imaginative potentials works.

Similar to concepts of 'space', which do not only contribute to shape our understanding of the world but also limit our possibilities to act and locate us within a more or less coherent structure of rules and rule, concepts of 'time' give meaning to social practices just as social practices make 'time' become a social reality. In his contribution *Hogne Øian* focuses on social practices of work and consumption as co-constitutive for the existence of linear time as a cognitive and ideological category. He relates the concept of linear time to the concept of a professional career and the temporal structure of everyday life in modern society. Consequently, becoming unemployed means on the one hand to become de-located from the dominant means of giving meaning and direction to one's life. On the other hand, being unemployed means to be cut off from the most obvious reference to (linear) time while clock time remains to be highly significant to conceptualise social life.

The relationship between a standardised linear time and the possibilities and limits of agency, discussed in Øian's contribution, is revisited by *John Murphy*, *Steven Arxer* and *Linda Belgrave* with regard to the ostensibly inescapable necessities of globalisation and its pre-given time structure. In their text, Newtonian time serves, among other things, as an evidence for the inevitability of globalisation and as the background for arguments limiting people's choices according to a distinct rationality. In order to re-conceptualise possibilities of agency as a collective praxis to alter the existing economic order the authors advance alternative understandings of 'time' as well as their social implications.

Under the heading *Agency and Resistance*, a second group of essays engages with the possibilities of the change of social arrangements by questioning modes of inclusion and exclusion, of the formation or the construction of identity as well as the production of knowledge.

Agata Stopińska's contribution focuses on changing gender roles in Poland and Ireland. Starting from Butler's concept of performativity and the fluidity of identity, she takes a look at the changing concepts of femininity and masculinity by looking at linguistic realisations of these concepts as well as mapping changes in gender issues both in Poland and Ireland by taking into account the cultural and socio-political contexts. Stopińska detects profound changes in the way gender is done as essentialist views seem to be much more difficult to uphold.

Orly Benjamin and Sarit Sambol explore the effects of travelling forms of feminism on local social structures and biographies. Taking as their point of departure a feminist economic empowerment project, they try to analyse the positive effects transnational feminist endeavours might yield. Benjamin and Sambol come to the conclusion that feminist projects may be localised by an enhanced spirit of women supporting each other but they may also loose some of their spirit in their encounter with local forms of power and poverty.

Agnieszka Kołodziej-Durnaś discusses major changes of the political and social system in recent years in Poland. She considers Poland after 1989 to be a special case in terms of changes within structures of self-identities and 'designing' biographies, influenced not only by high modernity or postmodern times as well as globalisation, but also by system transformation mechanisms. Kołodziej-Durnaś discusses ideas of the reflexive monitoring of action, agency and life politics, and reviews Polish students' reflexivity and agency in a broader cultural context of contemporary social changes.

While the increased use of surveillance systems, crime-fighting policies and endless consumer profiles seem to call for ever more sophisticated forms of identification, *Catarina Frois* contrasts these with a trend towards anonymity. She argues that the need for identification and the need for anonymity are not mutually exclusive but rather interdependent ways of interaction and communication within society and together form an important part of an individual's reality as a social being. The active embrace of anonymity serves as a means to escape control even if it is limited in time and space.

If knowledge production is constitutive for our understanding of social reality and in this sense essential for the implications of agency as well as the possibilities of resistance against and altering of social structures and institutions, than the text of *Steven Corcoran* is important twofold. First, it delineates philosophy as a means of the production of knowledge and second, it discusses the knowledge which is produced by the philosophy of Jacques Rancière. Corcoran analyses Rancière's work on the conception of political subjectivity by examining the links between philosophy and politics. Doing so, he is able to show the performative power of philosophy as well as the critical potential and limits of Rancière's practice of philosophy.

The third chapter, *Aesthetic Representations*, deals with philosophical and literary representations of agency which are firmly located in various social realities.

Dirk Wiemann provides a reading of two paradigmatic Restoration texts, Hobbe's "Behemoth" and Milton's "Paradise Lost", as widely discrepant interventions into the contest over agency and political enfranchisement which remained largely unresolved after the English Revolution. While both texts attempt to articulate the religious master code by means of a modern visual regime of linear perspective with vision being thematised as a privileged metonymy of agency, Hobbes and Milton arrive at very different conclusions where the struggle over agency is concerned.

Jatin Waglé explores the question of resistance of and in discourses by turning to the works of Theodor W. Adorno. He addresses the praxis of resistant presentational forms in Adorno's complex *oeuvre* by mapping his theory of discursive resistance, which is historically and discursively contextualised, as well as analysing the English language reception of Adorno's multifaceted writings. Waglé shows the translations to have become a hybridised embodiment of diverse critical subjectivities and thus a profoundly political act.

The paper of *Leena Petersen* concentrates on the representation of 'time' and 'space' as a historical concept of culture and social reality. Through a reading of Walter Benjamin's unfinished text of the "Arcades Project", Petersen deals with 'time' and 'space' as essential elements of Benjamin's concept of allegory. Here, allegorical consideration can be understood as cultural critique and alternative historiography. Images of space and 'time' as well as methods and modes of their visualisation serve as benchmarks for our understanding of reality and, consequently, for a critique of culture. In Benjamin's work this is exemplified by among others elements of architecture, fashion, panorama, modes of lighting, the mirror or the figure of the *flâneur*.

Athenian drama as an instrument of civic identity is the central theme of *Rocío Orsi Portalo*'s paper. She regards it as a means of self-celebration of the polis, as a tool for self-indulgence and as an instrument to accomplish a critique on values and conventions of the Athenian democracy. Orsi applies first Esposito's approach to the field of the symbolic reproduction of social life and its relation to political agency showing how the representation of violence accomplished by Athenian tragedy can contribute to social catharsis, the elimination of violence in today's societies. Finally, she reflects on opportunities of reaching a deeper self-knowledge than the one achieved by means of reflexive or philosophical resources alone.

Ulf Schulenberg's contribution analyses the ideological and rhetorical complexity of the U.S. implied to be a total liberal democratic space that pretends to be governed by a sort of consensus which in fact leaves no room for voices of resistance and rhetorical forms of dissent. Schulenberg discusses the question of whether poetic agency is capable of effectively confronting the hegemony of liberal democracy in the U.S. Firstly, he scrutinises Richard Rorty's conception of a literary culture; secondly, he focuses on two authors of the American Renaissance: Ralph Waldo Emerson and Walt Whitman. Conclusively Schulenberg identifies the creative and innovative work of the American strong poet as the one that can play a crucial role in the creation of the possibility of a counterhegemony.

The last set of essays, subsumed under the title *Agency and Its Spatio-Temporal Dimensions*, deals with agency – as a prerequisite for any kind of social praxis – which needs to be looked at as a concept embedded into time and space. In this sense any analysis of certain social institutions and arrangements as well as any attempt to change them needs to address spatial and temporal dimensions.

In his paper *Holger Rossow* deals with the radical restructuring of the world capitalist system into a regime of more flexible systems of production, distribution and consumption. Within this 'new world', temporal and spatial distances seem, on the one hand, to have lost their relevance and, on the other – this 'new world' is polarising rather than homogenising the human condition. Consequently, the possibilities of agency seem to have changed and traditional forms of political action are becoming undermined. But at the same time, as Rossow outlines, there are new opportunities to re-claim agency.

Tarja-Liisa Laaksonen's paper deals with current discourses of organisational culture as they are expressed in organisation studies and the social sciences. She scrutinises the complex interplay of space and agency via both domination and human agency, the influence of political globalisation issues, and emphasises the importance of local and cultural aspects in organisational life. She reflects on some clusters of major discursive themes indicating democratic deficits and revisits the organisational culture discourses which create both appropriate and 'excellent' cultures, and notions of the self-regulating and entrepreneurial self in various institutional environments. Laaksonen looks closer at discourses focused on fixed subjectivities, control over organisational subjects, as well as the domination of entrepreneurial discourses over workers' subjectivities.

Jörg Meyer explores the way in which agency gets lost in contemporary stories about world politics as told by the discipline of International Relations. Instead, a bulk of seemingly "testable" hypotheses is produced which are designed for problem management in the "real world" and result from an aspiration to scientific respectability. By using the example of the current preoccupation with norms in IR, Meyer shows that the routine acknowledgment of a capacity to initiate or to refuse vanishes in favour of causal explanations which foreclose the possibility of agency.

The contributions collected in this volume are, obviously, not intended to supply ready-made solutions to really (un)tie the knot binding 'time', 'space' and 'agency' but rather hope to stimulate the ongoing discussions by providing new threads.

alternative times

and postmodern mapping

Dagmar Reichardt

Mapping Sicily

From Postcoloniality to Neo-Metaphorization

Where and how do you map Sicily? For those who think that they readily have a clear picture of the island, most of the time knowing Sicily primarily means looking up Sicily in an atlas. Demarcating or imagining a (foreign) territory in picture form, i.e. on a map, is a process with a long tradition dating back to 1585, when Gerhard Mercator published his extensive *Atlas, sive cosmographiae meditationes de fabrica mundi.*[1] That procedure constantly pursues the (distant) goal – even when only mentally – of taking hold of a certain area, of occupying it, conquering it, capturing it for oneself, exploiting it and/or to dominate it as well.

Yet Sicily can also look like a woman who becomes an allegory for the *Regnum Siciliae*.[2] This is the case in a depiction that dates back to 1640. It was drawn up by the renowned Dutch cartographer Johannes Jansson, who publicized the figure that same year in a French compilation of maps bearing the title *Le nouveau théâtre du monde ou Nouvel Atlas*.[3] Here the land receives a likeness, it's being personified. It not only has outlines, it has an incarnate shape with a three-dimensional effect. I'll return to this body of a woman later on.

So there are two ways of imagining a place on this globe: mathematically-spatially, or three-dimensionally-humanly. I call these principles of human fantasy occupying/power-oriented versus symbolic/metaphoric. I would like to use the example of Sicilian cartography in the following to illustrate how the op-

[1] The notion of *atlas* derives from this title published on the frontispiece of his work (cf. Dufour/La Gumina 1998: 38/39) and defines in the 17th century a domain over which the Dutchman held the unchallenged supremacy. Schneider (2004: 54) indicates 1595 as year of publication of Mercator's *Atlas*. Dufour/La Gumina (1998: 38) distinguish between maps of Sicily that originate from an atlas and such that were published as loose leaves.

[2] The representation of a land, state or continent as a woman has a long tradition in the history of maps (cf. Ueckmann 2004).

[3] The frontispiece of the French edition from 1640 shows, among other things, allegorical representations of the continents.

fig. 1: Johannes Jansson: *Sicilia Regnum*, 1640 [detail]

pressive status of postcoloniality on the one hand and the opposing tendency towards metaphorization on the other have characterized the image that we associate with Sicily today. Before delving into that, I am going to deal briefly with Sicily's postcolonial status, a point closely linked to the island's history, so that I can follow up with a more in-depth examination of the historic backgrounds of cartographic development. A concluding step is to assess the value of the cultural representation emerging from the maps. This is intended towards answering the question – To what extent can Sicily together with Leonardo Sciascia be regarded as a metaphor for Italy, Europe or even the whole world from a postcolonialist point of view?

Postcoloniality

"Eppure da duemila cinquecento anni siamo colonia" (Tomasi di Lampedusa 1996: 170) – "And yet we have been a colony for two thousand, five hundred years": In Tomasi di Lampedusa's world-famous novel *Il Gattopardo* (1958; Engl. *The Leopard*), this is how Sicilian nobleman Don Fabrizio Salina sums up the problems Sicily had to face in the course of Italy's unification process around 1860. As a matter of fact, when relying on Tomasi's account one can say that Sicily has been forced to lead an existence as a "colony" for what has meanwhile become 2,600 years, since the post-Hellenistic era began cultivating a postcolonialism *avant la lettre*. The so-called *Continente Sicilia* had been colonized by the Greeks since the beginning of the 8th century BC,[4] and experienced its last foreign invasion under the rule of the Spanish Bourbon dynasty during the 19th century. Nonetheless, postcolonialist discourses on the island's society, politics and culture continue to have an effect to the present day. Viewed from this standpoint one can say that the island is in a permanent state of decolonization, involved in other words with the renunciation of both foreign invaders and external influences on power.

As I have already explained elsewhere (cf. Reichardt 2006a; Reichardt 2006b: 92), and even though I cannot go into more detail here, due to Sicily's special nature it seems to me to be appropriate in this case to principally refer less to *postcolonialism* as a topic, and instead to speak rather more of real *postcoloniality* on the part of Sicily.

History of Cartography

Let's move on to the history of cartography, particularly Sicilian cartography: To all intents and purposes, the historic development of the art of map-making is marked by the functional change away from the historically divine *Mappae mundi* – maps drawn up in monasteries offering orientation in a realm historically mandated by God's saving grace – to the topographic maps of modern times (cf. Schneider 2004). Among other aspects, these have formed the basis for references to science and for using geography as a spatial science. Since then, such things as satellite photos or orientational guides via GPS (*Global Positioning System*) are being measured in terms of their purported realistic correctness and the degree of their geographic precision.

A clay tablet from Mesopotamia is regarded as one of the oldest maps of the world handed down over the centuries. In the 6th century BC, Babylonians etched the world into it as a circle with Babylon as its center. Analogously, the drafter of the Ebstorfer World Map dating back to the 13th or 14th century AD shifts Jerusalem into the center of his manuscript: Only two cities are emphasized through the symbol of a flag – Jerusalem and Lüneburg in today's Germany. In accord with the message of this map, the royal house of the Lüneburger Welfen ought to promptly set off on a crusade to liberate Jerusalem (to the east). In the drafter's view, the Holy City is midpoint for the world, a

[4] The first wave of Sicily's colonialization by the Greeks lasted until 688 BC when a mixed group of Rhodesians and Cretans founded the Sicilian colonial town of Gela (cf. Finley 1994: 38/39).

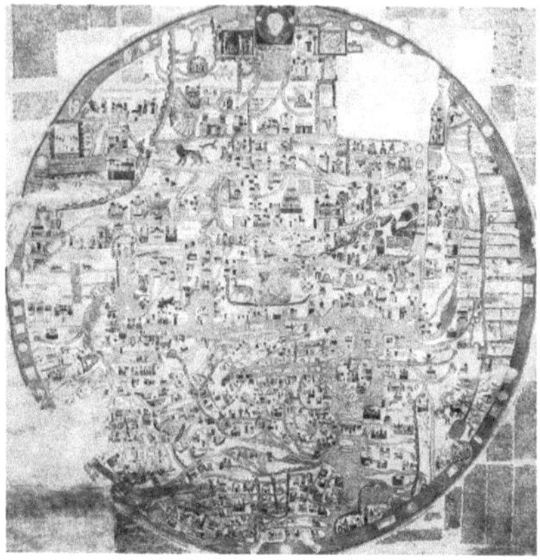

fig. 2: *Ebstorfer Weltkarte*, 13th or 14th century AD [total view]

place where order reigns. Chaos lies in wait at its margins, symbolized by the man-eating Gog and Magog to the north, and by birdmen or humans without ears to the south. Sicily too is clearly recorded to the southeast in the shape of a heart, the precise contours of the island still being unknown.

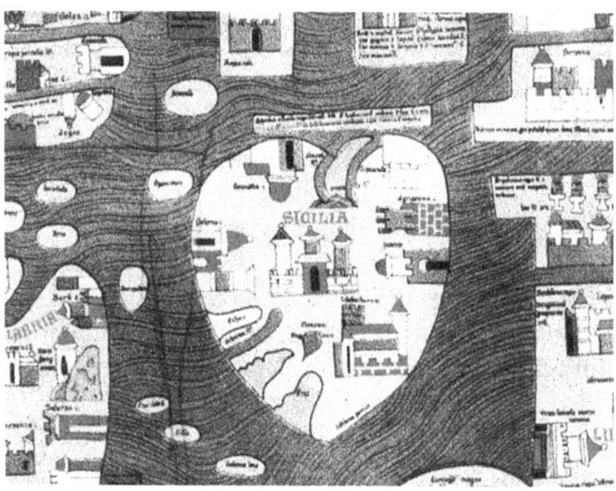

fig. 3: *Ebstorfer Weltkarte*, 13th or 14th century AD [detail: Sicily]

In other words, medieval maps had a dogmatic-religious background (representation of the power of the church) or pursued a theological-philosophical finality. Seen in a formal sense, they primarily possessed a mythical, symbolic value. Although Sicily does figure as a bridge between Africa and Italy in such medieval documents, i.e. as a strategic base for Byzantium, the spatial mapping and drawn profile accorded to the isle belong without a doubt in the realm of fantasy (cf. Dufour/La Gumina 1998: 15).

Maps of Sicily

As regards the historic development of maps of Sicily, that development is to be considered within this general framework: The first printed map of Sicily reiterates a draft of Sicily and Sardinia drawn up by Ptolemy (i.e. the Alexandrinian-Greek geographer Claudius Ptolemaios, approx. 100-180 AD),[5] and dates back to 1478.

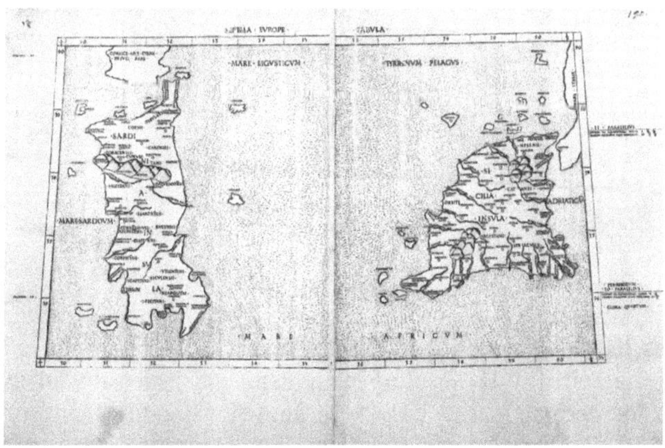

fig. 4: Claudius Ptolemaios: *Septima Europe Tabula*, 1478 [on the right: *Sicilia Insula*]

On Sicily that map was considered to be just as much a "bestseller" as the map of Sicily by Sebastian Münster that followed. Münster's map was published in his *Cosmographiae Universalis* in 1550 (cf. Dufour/La Gumina 1998: 36).
During the Renaissance of the 15th and 16th centuries, the rebirth of geography from ancient times builds upon the fundamental Ptolemaic period (as is the case with Münster). The focal point for this work shifted slowly but surely from the center of Flemish cartography (Abraham Ortelius and Gerhard Mercator) in Antwerp to Leiden and Amsterdam, where Dutch production assumed leadership. After the invention of letter-press printing in the 15th century, maps circulated at first in the form of xylographies (woodcuts), then as copperplate engravings (cf. Dufour/La Gumina 1998: 36). These production techniques were

[5] Ptolemy's world system considered the earth to be the center of the motions of the sun, moon and all other planets. Until the Copernican shift at the beginning of the 16th century which fixed the sun as such a centre, Ptolemy's major work entitled *Almagest* formed the basis of the astronomical knowledge.

fig. 5: Sebastian Münster: *De Italia Siciliae Insulae [...]*, in: id.: *Cosmographiae Universalis*, 1550

replaced by lithography during the *Seicento* (17th century). Cartography now experiences its Golden Age, a time in which it is increasingly regarded as a science in the spirit of 18th-century enlightenment, especially in England and France: What became particularly well-known were the maps of Sicily by Frenchman Guillaume Delisle, the map by Austrian General Samuel von Schmettau (who submitted a geographically individualized map of Sicily in 1809/10 on the orders of Emperor Karl VI), and the 1823 *Map of Sicily* by British Royal Navy Captain William Henry Smyth.

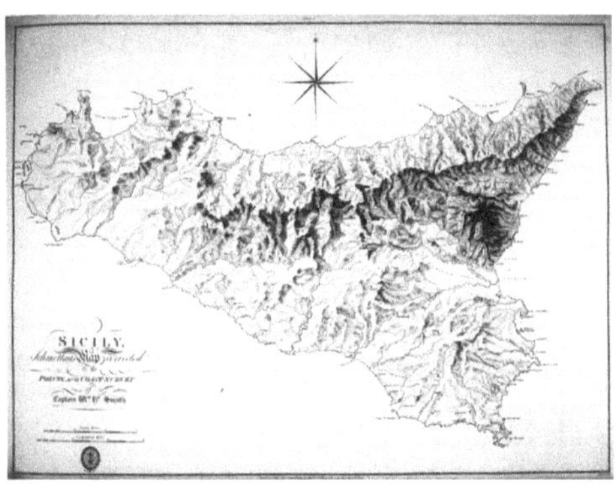

fig. 6: William Henry Smyth: *Sicily, Schmettau's Map Corrected [...]*, in: id.: *The Hydrography of Sicily, Malta and the Adjacent Islands*, London, 1823

During the 19th century, land maps eventually served more and more for purposes of warfare and military personnel. With the founding of the Topographic Institute in Palermo in 1808 (cf. Dufour/La Gumina 1998: 34), the new land map of Sicily from 1866 which had been produced in Italy, and as a result of the emerging improvement of surveying techniques, this situation rang in the end of historically drawn maps of Sicily (cf. Dufour/La Gumina 1998: 5).

Cultural Representation

What can we deduce? – As Liliane Dufour (cf. Dufour/La Gumina 1998: 42) concludes in her article from 1998 on the "Imago Siciliae", even today maps represent a never-achievable approximation of the island's reality, and thus an infinite *work in progress* phenomenon posing a basic postmodern principle *par excellence*. Even a satellite photo freezes only one of infinitely many potential moments. What is symbolized thereby is the postmodern view oriented on plurality, i.e. that there is not an absolute reality unto itself, but many realities in their plural instead. So, in the end maps themselves are mostly not any form of text, yet they are a weighty means towards causing the readers to write a text of their own or towards leading them to art. The maps thus take on an ambivalent, hybrid position as hinge or interface between reception and creation. The favorizing of the visual corresponds completely with the thesis of the *iconic turn*.[6] According to that thesis and pursuant to Horst Bredekamp, that turn marks our Western societies of today as a cultural shift from text to image, and in turn tills a fertile, receptive soil for the map genre (cf. Burda/Maar 2004: 15).

However, no map can make do without text, and maps are frequently the basis for texts: whether literary or academic texts, speeches or political agreements, journalistic or didactic commentaries etc. (cf. Buzan 1993). Our rapid study of the history of Sicilian cartography demonstrates clearly that two moments have made their mark on it: an empirical recording on the one hand, i.e. mapping bound to reality, and the cultural representation of the respectively prevailing view of the world (*Weltbild*) on the other. As Nicholas Visscher's world map from the end of the 17th century shows, a complete picture is given only through the interplay of geographical calculation and fantasy-filled

[6] I am taking over the notion of *iconic turn* from the programmatic title of a transdisciplinary lectures series held at the University of Munich in the period ranging from summer 2002 to summer 2003. The series dealed with the increasing importance of visual expressions within social communication in the fields of medicine and culture, i.e. science and humanities. The lectures were published under the same title (Burda/Maar 2004). In the introduction to this book the German art historian Horst Bredekamp explains how the formulas of a *pictorial* or *iconic turn* derive from the notion of the *linguistic turn* (cf. Burda/Maar 2004: 15). For further details on Visual Culture Studies cf. the basic texts by Mitchell (1994), Boehm (1994) and Mirzoeff (1999).

allegorization (in the four corners we see allegories for the four elements: fire, air, water, earth). At this point I am going to return to the woman's figure on Jansson's map of Sicily from 1640 which I had mentioned at the beginning. The woman embodies the Regnum Sicilia. Her position as the frame for a cartographic description of Sicily is necessary to ensure completeness with respect to what holds the world and/or Sicily together: She stands for what the cartographer is withholding, does not understand how to say in a different way or simply doesn't know.

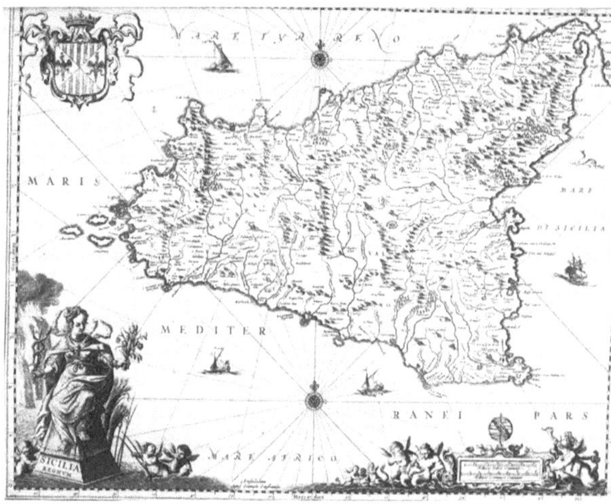

fig. 7: Johannes Jansson: *Sicilia Regnum*, 1640 [total view]

Accordingly, the value of Sicily's cartography has decisively been culturally marked. In the concepts used by *Italian Cultural Studies* (Forgacs/Lumley 1996), both the "Imagined Italies"[7] and the "Images of the South"[8] belong to the topic of examination for academic analyses. Even as late as the 1990s, a cartoon drawn by Giorgio Forattini for the September 18th, 1990 issue of *La Repubblica* displays an example of pejorative symbolization in relation to Italy's southernmost island[9] (cf. Forgacs/Lumley 1996: 103): On the front page of the Italian newspaper appears an intentionally amputated map of Italy which is not only reduced to the *Mezzogiorno* (Southern Italy) but also upside down when compared to present-day cartographic conventions. Forattini shaded three regions (Campania, Calabria, and Sicily) where the problem of orgnized crime

[7] Forgacs/Lumley 1996: 13-17 (chapter by John Dickie).
[8] Forgacs/Lumley 1996: 72-87 (chapter by Gabriella Gribaudi).
[9] The south of Sicily located islands *Isola di Pantelleria*, *Pelagian Islands* and *Isola di Lampedusa* are Italian State territory as well.

(Mafia and Camorra) is at its most serious. A warning sign in the middle of the drawing reads "Attenzione!!! State entrando nella zona dove si ammazzano i bambini" – "Beware!!! You are entering the area where they kill children". This "fantasy map" or "imaginary map" (Forgacs/Lumley 1996: 102) shows that the symbolic depiction of the world definitely did not become lost during the Middle Ages:[10] Sicily, Italy's south and the copy "blurb" stand for the semantic sectors of the Mafia, crime, illness, exclusion etc. "Mappa means depiction, thus Mappa Mundi means about the same as depiction of the world" (Schneider 2004: 25).

fig. 8: *Carta simbolico-geografica dell'Italia irredenta*, approx. 1913

In light of this "Symbolic-geographical map of Italy" (fig. 8) for example, that meaning could be extended to a depiction of a world view (*Weltanschauung*), whereby its symbols, framing and coloring recall the "unredeemed" regions of Trento and Trieste. The map expresses a political discourse by noting Italy's situation prior to the First World War: The oversized *Italia* in the middle of the picture leads a group of soldiers (including mountain troops with Tyrolian hat and feather, the so-called *Alpini*) marching toward two women chained and bound to the stake symbolizing Trento and Trieste. The map is framed by

[10] As references for this thesis Ute Schneider produces e.g. the *Clark's Chart* from 1822 and the *Weltatlas für Kinder* from 2004 (cf. Schneider 2004: 106/107 and 55).

portraits of heroes from the *Risorgimento* i.e. the 19th century (Garibaldi, Mazzini, Vittorio Emanuele II, Cavour), so that the map's political and theoretical aspects in terms of power, which Ute Schneider (cf. Schneider 141) points out in her publication on the "Power of Maps" – *Die Macht der Karten* (2004) – cannot be overlooked.

Neo-Metaphorization

In an age of the accomplished territorial mapping of Sicily, the value of the space's cultural representation is meanwhile on the rise: All of the major stages of occidental history are reflected in Sicily's eventful historic past. Often enough, the dynamics of Sicily's own self-determination were representative of leading intellectual events going on in the world. Sciascia's thesis of a *Sicily As Metaphor* (Sciascia 1994) is to be understood in this sense:[11] Sicily actually is a metaphor for Italy, Europe, yes, even for the whole world. And making that thought comprehensible can hardly be done more impressively and illustratively in any other field than in cartography. In Sicily's history, the figurative power of man's imagination plays a role that marks its culture and forms the basis of science and literature right from the start.

In more recent times, after Sciascia, an increasing interest in a neo-metaphorization has been established, particularly in academic terminology since the 1980s.[12] For instance, iconic articulations can be found in Deleuze and

---

[11] The original Italian title *La Sicilia come metafora* was published in 1979. A few months before appearing on the Italian book market the volume was originally printed and entitled in French language: *La Sicile comme métaphore. Conversations en italien avec Marcelle Padovani* (Paris: Editions Stock), Sciascia's interviewer Padovani being a correspondent of the French newspaper *Le Nouvel Observateur*. James Marcus translated the book into American English in 1994 (Sciascia 1994).

[12] Of course the phenomenon is verifiable also before the 1980s. Concerning the cultural terminology in the 20th century, metaphorical adoptions can be quoted also outside Literary Studies, in the scientific fields of botany (see the concept of a *rhizome* by Deleuze/Guattari), biology (*Culturas hibridas* by Néstor García Canclini), linguistics (*Creolizzare l'Europa* by Armando Gnisci), ethnology (*Traité de nomadologie [...]* in: Deleuze/Guattari 1980: 434-527), medicine (the *hétérotopies* by Michel Foucault), physics (Stephen Greenblatt's idea of the circulation of social energy), history (the *Habsburg* theory in Feichtinger/Prutsch/Csáky 2003) or on the industrial sector of electronics i.e. in the IT branch (see the metaphor of a net or the cultural *Netzwerk-Design* in Welsch 2002). The theoretical models of an *Empire* (Hardt/Negri) or *mapping the world* (cf. the metaphor of a "carte" which is inspired by cartography in Deleuze/Guattari 1980: 20) also join the tradition of an increasing displacement towards visual terms in the 20th century. The latter is ascertainable since the appearance of Vasconcelos' book entitled *Raza cósmica* in 1927, but especially after the end of World War II, when a political forming of concepts followed which refered to metaphors such as the *Iron Curtain*, the *Cold War* or the *Third World*.

Guattari, who drafted first *mental maps* using their metaphor of a "carte" (Deleuze/Guattari 1980: 20), or through the cartography in their *Mille plateaux* (1980). On a theoretical level of reception, Deleuze's and Guattari's metaphorics of the "carte" were developed further into a postcolonialist concept of *mapping the world*. Among other works, this led to Homi K. Bhabha's *Third Space* metaphor (Bhabha 1998), as well as to the metaphoric recoding of the motto *When margins become the center* (Adobati et al. 2001).

Alongside satellite photos, X-rays or a glance into the nano-world through a scanning tunnel microscope (cf. e.g. Burda/Maar 2004: 129 pp.; Harmon 2004: 14 pp.), "personal geographies" (Harmon 2004) and the previously mentioned *mental maps* – i.e. the representation of space in our imaginary world (the term goes back to Tony Buzan; cf. Buzan 1993) – are increasingly becoming the focus. Among other areas, the link to literature and language is to be mapped in the conceptual pre-drafting of a narrative *Histoire* before the author's inner eye, or in the reader's imagination that accompanies the reading of a text. As I explained elsewhere, the *mental map* of *sicilianità* ("Sicilianness") poses the true main subject matter, especially on the collective level, around which all narrative texts by Sicilian authors and writers revolve (cf. Reichardt 2004).

Without a graphic vividness, language – even literary language – cannot be written or brought to life, let alone grasped in a comprehensible way that is understood or put to gainful use. The development of Sicilian cartography shows in an exemplary manner that, when viewed iconographically, it is just as multilayered as the cultural history of this South European island. It is the *mappa*, in other words the cartography of cultural influences that Sicily has experienced and put into practice, which makes up this region's authenticity. And in the process it emerged that the power of maps is not least of all an aesthetic power.[13]

[13] Cf. Knipp 2004. Knipp emphasizes not only the aesthetic significance of geographical maps but also stresses the political function of cartography: Stalin parceled out Poland on 28-09-1939 bending over a map, holding a pencil in his hand, European colonizers drew up the inland frontiers of Africa with a ruler, and in the pact of Tordesillas, in 1494 Spaniards and Portuguese agreed upon the southwestern distribution of the landmasses laying in the just "discovered" America, without even knowing their realistic dimensions of the landmasses. In this sense, possessing a map made of paper often implies the possession of the corresponding land in reality. So, with the help of maps, men not only depict their world but recreate it at the same time and arrange it according to their own image.

Works Cited

Adobati, Chantal, Maria Aldouri-Lauber, Manuela Hager & Reinhart Hosch, ed. (2001). *Wenn Ränder Mitte werden. Zivilisation, Literatur und Sprache im interkulturellen Kontext. Festschrift für F. Peter Kirsch zum 60. Geburtstag*. Wien: WUV Universitätsverlag.

Bhabha, Homi K. (1998 [1994]). *The Location of Culture*. London & New York: Routledge.

Boehm, Gottfried (1994) "Die Wiederkehr der Bilder". *Was ist ein Bild?* Ed. Gottfried Boehm. München: Fink, 11-38.

Burda, Hubert & Christa Maar, ed. (2004). *Iconic Turn. Die neue Macht der Bilder*. Köln: DuMont.

Buzan, Tony (1993). *The Mind Map Book*. London: BBC Books.

Deleuze, Gilles & Félix Guattari (1980). *Mille Plateaux. Capitalisme et schizophrénie*. Paris: Minuit.

Dufour, Liliane & Antonio La Gumina, ed. (1998). *Imago Siciliae. Cartografia storica della Sicilia 1420-1860*. Intr. Giuseppe Giarrizzo. Catania: Sanfilippo.

Feichtinger, Johannes, Ursula Prutsch & Moritz Csáky, ed. (2003). *Habsburg postcolonial. Machtstrukturen und kollektives Gedächtnis* (Gedächtnis – Erinnerung – Identität vol. 2). Innsbruck: Studien Verlag.

Finley, Moses I. (1994^2 [1968]). *Das antike Sizilien. Von der Vorgeschichte bis zur arabischen Eroberung*. München: Deutscher Taschenbuch Verlag.

Forgacs, David & Robert Lumley, ed. (1996). *Italian Cultural Studies. An Introduction*. Oxford: Oxford UP.

Harmon, Katharine, ed. (2004). *You Are Here. Personal Geographies and Other Maps of the Imagination*. New York: Princeton Architectural Press.

Knipp, Kersten (2004). "Das historische Buch: Macht und Ästhetik der Karten" [review]. *Neue Zürcher Zeitung*, 27.10.2004.

Mirzoeff, Nicholas (1999). *Introduction to Visual Culture*. New York: Routledge.

Mitchell, William J. T. (1994). "Pictorial Turn". *Picture Theory*. Ed. William J. T. Mitchell. Chicago: U of Chicago P, 11-34.

Reichardt, Dagmar (2004). "Concetti di sicilianità: Sciascia, Bufalino, Consolo, Camilleri, Freni e Grasso". *Lingue e letterature in contatto. Atti del XV Congresso dell'A.I.P.I. Brunico, 24-27 agosto 2002, vol. 2*. Ed. Bart Van der Bossche, Michel Bastiaensen & Corinna Salvadori Lonergan. Firenze: Cesati, 441-451.

Reichardt, Dagmar (2006a). "Italia ibrida. La Sicilia come terzo spazio nel discorso interculturale". *Italia e Europa: dalla cultura nazionale all'interculturalismo. Atti del XVI congresso A.I.P.I. Cracovia, agosto 2004*. Ed. Bart Van der Bossche, Michel Bastiaensen & Corinna Salvadori Lonergan. Firenze: Cesati [forthcoming].

Reichardt, Dagmar (2006b). "Paradigma mundi? Die Geschichte des postkolonialen Siziliendiskurses zwischen literarischer Alterität und Identität". *L'Europa che comincia e finisce: la Sicilia. Approcci transculturali alla letteratura siciliana. Beiträge zur transkulturellen Annäherung an die sizilianische Literatur. Contributions to a Transcultural Approach to Sicilian Literature*. Ed. and with a preface by Dagmar Reichardt, with the collaboration of Anis Memon, Giovanni Nicoli & Ivana Paonessa, (Italien in Geschichte und Gegenwart vol. 25). Frankfurt a. M., Berlin, Bern et al.: Peter Lang, 87-107.

Sciascia, Leonardo (1994 [1979]). *Sicily As Metaphor*. Tr. James Marcus. Marlboro, Vt.: Marlboro Press.

Schneider, Ute (2004). *Die Macht der Karten. Eine Geschichte der Kartographie vom Mittelalter bis heute*. Darmstadt: Primus Verlag.

Tomasi di Lampedusa, Giuseppe (1996 [1958]). *Il Gattopardo. Opere*. Ed. Giuseppe Tomasi di Lampedusa, intr. Gioacchino Lanza Tomasi. Milano: Mondadori, 19-318.

Ueckmann, Natascha (2004). "Voyages en Orient au féminin – un discours marginalisé". *Etudes féminines/gender studies en littérature en France et en Allemagne. Actes du colloque organisé par le Frankreich-Zentrum de l'université Albert-Ludwig de Fribourg, le 6 et 7 décembre 2002*. Ed. Rotraud von Kulessa. Freiburg i.B.: Frankreich-Zentrum, 123-143.

Welsch, Wolfgang (2002). "Netzdesign der Kulturen". *Zeitschrift für Kulturaustausch* 1 [special issue: "Dialogue with the Islam. Between Right and Reality"]. <www.ifa.de//zfk/themen/02_1_islam/dwelsch.htm> (22-03-2006).

Illustrations

Fig. 1: Doufour, Liliane / Antonio La Gumina (ed.) (1998). *Imago Siciliae: Cartografia storica della Sicilia 1420-1860*. Catania: Sanfilippo, frontispiece.

Fig. 2: Schneider, Ute (2004). *Die Macht der Karten: Eine Geschichte der Kartographie vom Mittelalter bis heute*. Darmstadt: Primus Verlag, 24.

Fig. 3: Schneider, Ute (2004). *Die Macht der Karten: Eine Geschichte der Kartographie vom Mittelalter bis heute*. Darmstadt: Primus Verlag, 24 (detail).

Fig. 4: Doufour, Liliane / Antonio La Gumina (ed.) (1998). *Imago Siciliae: Cartografia storica della Sicilia 1420-1860*. Catania: Sanfilippo, 48.

Fig. 5: Doufour, Liliane / Antonio La Gumina (ed.) (1998). *Imago Siciliae: Cartografia storica della Sicilia 1420-1860*. Catania: Sanfilippo, 71.

Fig. 6: Doufour, Liliane / Antonio La Gumina (ed.) (1998). *Imago Siciliae: Cartografia storica della Sicilia 1420-1860*. Catania: Sanfilippo, 261.

Fig. 7: Doufour, Liliane / Antonio La Gumina (ed.) (1998). *Imago Siciliae: Cartografia storica della Sicilia 1420-1860*. Catania: Sanfilippo, 120.

Fig. 8: Schneider, Ute (2004). *Die Macht der Karten: Eine Geschichte der Kartographie vom Mittelalter bis heute*. Darmstadt: Primus Verlag, 141.

Maria Antonietta Mariani

Engraving Space

Maps, a *World-Body* Portrait

According to Foucault (1964) space is the most obsessive metaphor for language because space provides language with a way to create and reach out. Through its close relation with space, the map is also "metaphorized". In this article, we will explore the metaphorical power of maps. We can follow the way in which maps are used to relate and to disclose our space for our thought. Korzybski (1941) talks about the map-territory relation when the message is not the object it defines. Bateson (1972) uses Korzybski's metaphor: the map is not a territory. What does become part of the map is simply the difference: a difference in landscape, in altitude and so on. This difference is an abstraction. The mapmaker's interpretation reshapes the object and gives it other meanings. In doing so, the relation between mapmakers and territories comprises more than recognisable signs.

Space/memory

> Man is called by the ancient a world in miniature and certainly this name is well applied, for just as man is composed of earth, water, air and fire, so is the body of earth. If man has in him bones which are the support and armor of the flesh, the world has rocks which are the support of the earth; if man has in himself the sea of blood, in which the lungs rise and fall in breathing, so the body of the earth has its oceanic sea which also rises and falls every six hours for the world to breathe. If from the said sea of blood spring veins which go on ramifying throughout the human body, similarly the oceanic sea fills the body of the earth with numberless veins of water.
>
> (Leonardo da Vinci)

Just like the body, the map is tied to space organisation. In Ancient Egypt, the Osiris body is broken up and each part becomes a district icon. In some African areas, when the king dies the capital is shifted to another place. The king's body as well as the capital modify space, and their transfer signs space and engraves memory.

A cartography is a space of memory. In his film *Memento* (2002), filmmaker Nolan describes the story of a man who loses his memory, while his body becomes the place of his memory. Signs are drawn on his body to remember. His skin, just like a map, gathers and holds his memory.

Every place can be found on maps. They give an order to all the world's places, choosing a north, a south, an east and a west. According to Mary Douglass (1970) the space-time organisation derives from the exploitation of differences. If we follow differences, in/out, up/down, man/woman, with/ without then an order appears. The fence creates a sacred space. Maps keep the concept of the sacred threshold.

In Lars Von Triers' film, *Dogville*, (2003) the protagonist, Grace, is on the run from gangsters. The tale is shot entirely on a skeletal stage, where houses have no walls and the names of the streets are written in white paint on the wooden ground. Just like on a map. The film tells the story of Grace's stay in Dogville, at first they appreciate her generosity, but slowly look upon her as something else, an object of secret depravity. Stepping over the threshold costs the foreigner a high price. It is the unknown risk living inside the traveller, and nomads and migrating peoples.

Space/society

> Regardless of the words, it seems the melodic contour of the song described the nature of the land over which the song passes. So, if the Lizard Man were dragging his heels across the salt-pans of Lake Eyre, you could expect a succession of long flats, like Chopin's "Funeral March". If he were skipping up and down the MacDonnel escarpments, you'd have a series of arpeggios and glissandos, like Liszt's "Hungarian Rhapsodies". Certain phrases, certain combinations musical notes, are thought to describe the action of the Ancestor's feet. One phrase would say, 'Salt-pan'; another 'Creek-bed', 'Spinifex', 'Sand-hill', 'Mulga-scrub', 'Rock-face' and so forth. An expert song-man, by listening to their order of succession, would count how many times his hero crossed a river, or scaled a ridge-and be able to calculate where, and how far along, a Songline he was. 'He'd be able', said Arkady 'to hear a few bars and say, "This is Middle Bore" or "That is Oodnadatta"-where the Ancestor did X or Y or Z. 'So a musical phrase', I said, 'is a map reference?' 'Music', said Arkady, 'is a memory bank for finding one's way about the world'.
>
> (Bruce Chatwin 1987: 82)

Just as in a mythical story, maps keep a traveller and society reading the endless change of human beings. So what is the object/territory behind the map, and what is it like?

In the Medieval age, topography was mainly religious. The Holy Land was the destination of philosophical travels and maps signalled the importance of the sacred space too. On the other hand, an imaginative topography is drawn by Homer's stories. There were two kinds of maps: *T-O* maps and *Zonal Maps*. Geographical theories inherited from Greek Science and mythology divided the earth in five regions[1], the so-called *Macrobius's Zonal Map*[2] is oriented with the north at the top, which preserves the Hellenistic tradition of following the North Star. While *T-O* maps are diagrams that originate from the Romans and later developed with Christian concepts, they are oriented with the east at the top; the *O* represents the known world and its circumfluent ocean; the *T* is the axis of the Mediterranean crossed by the meridian from the Don in Russia to the Nile. With the east at the top, the viewer focuses on Jerusalem and the Garden of Eden. In this way, philosophical thoughts and the social importance of religions are shown. The Medieval age was therefore characterized by schematic *T-O* map-types from Isidore's manuscripts or zonal diagrams to Macrobius's manuscripts. During the centuries, Christian themes are combined with Islamic decorative motifs and forms. The late-fourteenth-century picture of the known world is enriched with real and imagined portraits, landscapes, fauna, heraldry, ships, caravans. Medieval maps represent Ancient Greek and Roman mythology, the Bible and tales of medieval travellers.

At that time, the world boundary was fixed and no one would have dared to transgress it. Distances and places enter maps only after several centuries, when travelling becomes a scientific experience: the process of comparison starts. Renaissance travel introduces the same/different opposition. The phrase "like in Europe", leads to an object inside one's mental space. The new world is the place of mythical times. The old one considers itself advanced. Space difference is also time-society difference. In the past centuries maps were inspired by texts. Then, maps were filled up with differences creating the *thick map*.

During the centuries, the play of differences provides us with new maps. The world appears deeply changed. Some countries have disappeared, some of them have changed their names and boundaries, others have created new countries. Behind the new map-lines, stories shifted axes and international political strategies lie.

---

[1] Arctic Circle, Tropic of Cancer, Equator, Tropic of Capricorn, Antarctic Circle.

[2] Ambrosius Macrobius, a late-Roman philosopher whose writings were popular during the Middle Ages. He is known for his *Commentary on the Dream of Scipio*, part of Cicero's *De Republica*, in which Macrobius included his geographical theories inherited from ancient Greek science and mythology.

Territory comes into the map that widens to gather all the elements. Then it will trasform them giving back innumerable fragments. If we look at Google maps[3]: every street, every corner and every building could be seen and recognised. Is looking at a map the same as looking at the territory? Isn't there any space for imagination behind images? The map is no longer different from the territory. No distinction exits between them. Everything could be soon seen. The territory becomes a map of its own. Baudrillard argues

> abstraction today is no longer that of the map, the double, the mirror or the concept. Simulation is no longer that of a territory, a referential being or a substance. It is the generation by models of a real without origin or reality: a hyperreal. The territory no longer precedes the map, nor survives it. Henceforth, it is the map that precedes the territory — precession of simulacra — it is the map that engenders the territory..." (Baudrillard 1998: 166).

Space/technology

> In that empire, the craft of cartography attained such perfection that the map of a single province covered the space of an entire city, and the map of the empire itself an entire province. In the course of time, these extensive maps were found somehow wanting, and so the college of cartographers evolved a map of the empire that was of the same scale as the empire and that coincided with it point for point. Less attentive to the study of cartography, succeeding generations came to judge a map of such magnitude cumbersome, and, not without irreverence, they abandoned it to the rigours of sun and rain. In the western deserts, tattered fragments of the map are still to be found, sheltering an occasional beast or beggar; in the whole nation, no other relic is left of the discipline of geography.
>
> (Borges 1998: 320)

Technology modifies the mapmaker's approach and the viewer's perception. On one hand, maps are filled up with new information, on the other hand, maps are multiplied, broken up and innumerable fragments created. More and more sophisticated instruments are used, but a clear choice about data compressing can be seen behind them. Using technologies the mapmaker images a muddle of possible routes.

[3] Consult: www.cybergeography.org/atlas/geographic.html, www.maps.com, www.nationalgeographic.com, www.nsg.org, www.orlan.net, www.un.org/depts/cartographic/english/htmain.htm, www.visibleearth.nasa.gov

Pictures and numbers come together in the production of the manipulated image. Satellite photographs are used to predict the future. They were used before the war in Iraq to verify Saddam Hussein's arms and, more recently, the Tsunami disaster in Asia. The way of looking has been deeply modified. It has become three-dimensional. Maps show the evolution of events: past, present, future. Reality movement is followed.

By introducing digital technology, spatial perception of the city has deeply changed. GPS, for example, is a satellite navigation system used for determining one's precise location and providing a highly accurate time reference almost anywhere on earth or in earth orbit. It is a network of satellites and ground computers around the world which processes data received via microwaves from satellites, to give precise information on the position of an object. The object is looked at as a numerical coordinate on an orbital scale. This allows for an exact and absolute positioning on the globe. Just like in Borges'tale, topographical representation has grown as huge as the territories it represents. With their potential for displaying each pictured element at colossal scale, computerized satellite monitoring technologies have allowed for a nearly one-to-one representation. In this way, anyone and anything in the world can be tracked.

In bringing a map, we take a previous sense of space that we will go through. To use a map is a differences-hunt. On standing at a crossroad we try to orient the map looking at the confluence between this red building, the square down there – what will be that dome behind? – and the shouting hawkers with their stalls spreading in the next street. Somewhere – this is on the map – there should be a bridge. Yes, it is, we can smell it… Perhaps, we should choose the main road – the straight line – to overcome our fear of getting lost and getting entangled in the "muddle-city".

If a system of best running choice exists, the way of thinking space and moving through it changes. The discovering process of territory is modified. It becomes more external. Some images drive us: images rather than songlines, looking at them rather than listening to or smelling them. They are collected in a software rather than an inner "memory bank". In some way, we are not involved in the muddle-city but over it, looking at it from the satellite perspective, sliding on the fluid space of the city surface. The technology offers us the way of discovering and choosing. The system underlines a comparison distance/obstacles/time. Neither sound nor smell nor colour melt the process of choice. Do they perhaps disappear all together with the unforeseeable and inevitable "mistake" of getting lost?

Bodyscape

> Le tue città non esistono. Forse non sono mai esistite. Per certo non esisteranno più. Perché ti trastulli con favole consolanti? So

> bene che il mio impero marcisce come un cadavere nella palude, il cui contagio appesta tanto i corvi che lo beccano quanto i bambù che crescono concimati dal suo liquame. Perché non mi parli di questo? Perché menti all'imperatore dei tartari, straniero?
>
> (Calvino 1993: 59)

Maps position the site, people, and objects they represent. But maps can also take a wider array of social, political and economical data into account: income, age, race, sexual orientations, religion, languages, health, consumption. Nowdays, more and more diagram-maps are used to relate a polarized world. Social, political and economic indexes are worked out and compared.

Le Monde diplomatique Atlas collects data information and reveals the differences. Space appears as an image of differences in consumption, differences in development index, and so on. Symbols and colours underline the distance of given resources and capabilities. The play of differences goes on: more and more maps gather together differences, or rather, they continue to multiply differences. In Koolhaas's maps, all indexes are joined and melted showing the A-Z World, providing unusually powerful matching/relation/ difference. Maps as an image of political and economical power. Some parts of the globe are shown in newspapers to narrate stories of conflicts. The map has been broken in order to contain all differences. It happens in the press[4], where maps break into pieces. We could think of the map as a representation of the *world-body*, its image becomes the sum of scattered fragments. Every piece is a strong image, just like the mouth, leg, or breast.

In this way, map shapes and signs remind us of body art works. *Bodyscape*. We can consider the art works by the French artist Orlan: her *Carnal Art* lying between disfiguration and figuration is an inscription in flesh, so the body becomes a "modified ready-made". One's own body cut open can be observed without suffering. A new mirror stage shows an outside/inner vision of one's own body. Using performance, video, multimedia and scientific techniques – surgery – Orlan shapes and reshapes her own body. She creates an endless real and virtual morphing. Her work aims to impact on memory, thereby suggesting that one's thoughts and the recollection upon which they rely can somehow be tangibly shaped. Orlan's self produces self-imagery.

---

[4] I collected and analysed a sample of international press images which use maps. An analysis of their pertaining article titles revealed that they often remind us of the body. For instance, the article title *Operation Deep Pockets* (The Economist, December 2004) recalls a coat, something someone wears, or the other one, *When deadly force bumps into hearts and minds* (The Economist, January 2005), the words "hearts" and "minds" and, before them, "deadly" remind us of the body. Finally, *L'Ombre de Mitch sur l'Amérique centrale* (Le Monde diplomatique, Fevrier 2005) clearly shows the image of Central America as a broken limb.

Merleau-Ponty says "the body proper embraces a philosophy of flesh as the visibility of the invisible" (Merleau- Ponty 1970: 197). That is the *intertwining*. The skin is Orlan's metaphor for the surface of being as interface between inside and outside. Using morphing software Orlan melts her own image into the eyes of Gerome's Psyche, the forehead of Leonardo's Mona Lisa, the chin of Botticelli's Venus, the nose of a School of Fontainebleau Diana and the mouth of Boucher's Europa. By computing machine she works on her medically altered post-operational face, shaping a composed visage and finally creating the sequential self-portrait. This underlines the idea of the body as corporeal landscape, according to Deleuze and Guattari[5] (1980), painting may use a landscape as a face, treating one like the other.

In the same way the post-modern body is broken up, while some parts of it are isolated and manipulated, the same is true of maps. We could paint the map as a representation of the *world-body*. The map becomes the body, the *map-world-body,* where someone writes what has to be remembered. The body space has now become the space body. We could talk about a *map-skin*. Like the body, the map is not natural: its signs, movements, and its appearance, all of them are significant. The media, an extremely powerful *semiotic engraver*, underlines the body-space relation and the metaphorical strength of maps. The body image returns in what I would define *the post-human map*. It is a fragment that shows relations with missing pieces and more. The break in the map interrupts the association flow, causing a shock effect. The viewer perception is modified by differences multiplying and scattering.

Works Cited

Appadurai, Arjun (1996). *Modernity at Large: Cultural Dimensions of Globalization*. Minneapolis & London: University of Minnesota Press.

Baudrillard, Jean (1998). "Simulacra and Simulations." *Selected Writings*. Ed. Mark Poster. Stanford UP, 166-184.

Bateson, Gregory (1972). *Steps to an Ecology of Mind*. San Francisco, Scranton, London & Toronto: Chandler Publishing Company.

Bauman, Zygmunt (1999). *In Search of Politics*. Cambridge: Polity Press.

Benjamin, Walter (1955). *Das Kunstwerk im Zeitalter seiner technischen Reproduzierbarkeit*. Frankfurt am Main: Suhrkamp Verlag.

Borges, Jorge Luis (1998). *Collected Fictions*. New York: Penguin Putnam.

Calvino, Italo (1993). *Le città invisibili*. Milano: Mondadori.

Canovacci, Massimo (2004). *Sincretismi*. Roma: Costa&Nolan.

Chambers, Ian (2001). *Culture after Humanism*. London: Routledge.

Chatwin, Bruce (1987). *The Songlines*. New York: Viking Penguin.

[5] Deleuze & Guattari (1980).

Clifford, James (1986). *Writing Culture: Poetics and Politics of Ethnography.* Berkeley & Los Angeles: University of California Press.
Clifford, James & George E. Marcus (1988). *The Predicament of Culture: Twentieth-Century Ethnography, Literature and Art.* Cambridge, Mass & London: Harvard UP.
Da Vinci, Leonardo (1970). *The Notebooks of Leonardo da Vinci.* New York: Courier Dover Publications.
Deleuze, Gilles (1968). *Différence et Répétition.* Paris: Presses Universitaires de France.
Deleuze, Gilles (1996). *Divenire Molteplice.* (trans.) Verona: Ombre Corte.
Deleuze, Gilles & Fèlix Guattari (1980). *Mille Plateaux.* Paris: Minuit.
Dilke, O.A.W (1985). *Greek and Roman Maps.* London: Thomas & Hudson.
Douglass, Mary (1970). *Purity and Danger. An Analysis of Concepts of Pollution and Taboo.* Harmondsworth: Penguin.
Edson, Evelyn (1999). *Mapping Time and Space. How Medieval Mapmakers Viewed Their World.* London: British Library.
Foucault, Michel (1964). "Le langage de l'espace". *Critique* 203, 378-386.
Geertz, Clifford (1973). *The Interpretation of Cultures.* New York: Basic Books.
Gould, Peter & Rodney White (1986). *Mental Maps.* Boston: Allen & Unwin.
Harmon, Katharine (2004). *Personal Geographies and Other Maps of the Imagination.* New York: Princeton Architetectural Press.
Harvey, P.D.A. (1980). *The History of Topographical Maps: Symbols, Pictures and Surveys.* London: Thomas & Hudson.
Koolhaas, Rem (2003). *The Ultimate Atlas for the 21st Century.* Wired.
Korzybski, Alfred (1941). *Science and Sanity.* New York: Science Press.
Leed, Eric J. (1991). *The Mind of the Traveller. From Gilgamesh to Global Tourism.* New York: Basic Book.
Le Monde diplomatique (2003). *Atlante.* (trans.) Roma: Il Manifesto.
Lévi-Strauss, Claude (1955). *Tristes Tropiques.* Paris: Plon.
Mariani, M.A. (2005). "Visible and Invisible Worlds on Present Maps". *3rd International Conference on Globalisation and Resistance, Globalisation & Representation. Politics, Visuality, Performance, Construction, Reception.* Faculty of Arts and Architecture, School of Historical and Critical Studies. University of Brighton: March 12-13, 2005.
Mariani, M.A. (2005). "Bodyscape. The Post-Human Map". *(Re)Creating: Methodologies, Practices, Concepts Symposium Time, Space, Body, Image*, Goldsmiths College, University of London: September 1-3, 2005.
Merleau- Ponty, Maurice (1970). "Nature and Logos: The Human Body". *In Praise of Philosophy and Other Essays* Evanston: Northwestern UP, 197.
Penrose, Boies (1955). *Travel and Discovery in the Reinassance, 1420-1620.* Cambridge, Mass.: Harvard UP.
Presten, James (1972). *All Possible Worlds. A History of Geographical Ideas.* New York: Bobbs-Merril.
Remotti, Franco (1993). *Luoghi e Corpi. Antropologia dello Spazio, del Tempo e del Potere.* Torino: Bollati Boringhieri.

Filmography

Von Triers, Lars (2003). *Dogville.*
Nolan, C. (2002). *Memento.*

Hogne Øian

Linear time

The Enemy of the Unemployed

Introduction

Being out of work, one's everyday life ceases to be mastered by clock time. The reason is of course that it is the schedules of institutions like work and school that make the clock-time most relevant in everyday life. Considering the seemingly ever-increasing speed and lack of time in post-industrial societies, one could expect unemployment to be experienced as freedom from clock time. This could be so, at least in the Scandinavian welfare states, where unemployment only to a limited extent leads to an immediate economical marginalisation. Nevertheless, what rather seems to be the case is that clock time more than ever continues to cause frustrations. Before unemployment the problem was the constant relevance of clock time. Once out of work it is the lack of relevance of clock time that becomes a problem.

In the beginning of the 1990s there was a so-called crisis of unemployment in Norway that lasted for a few years only. With figures not much above 10 percent, the word crisis is of course an exaggeration. Compared to the situation in countries like Germany or Sweden, it is nothing special. Nevertheless, in Norway it was perceived like a crisis. The reason, I believe, is that the rather sudden increase in unemployment rates challenged the images most Norwegian have of their own nation; a nation secured by the abundances of the oil-economy and a well developed welfare state. During this period of rising unemployment I carried out anthropological fieldwork among unemployed youths in the suburbs of Oslo (Øian 2004). The method was mainly based on participant observation and interviews. The aim was primarily to find out how the unemployed youths coped with identity and morality in relation to time.

One simple fact became clear to me, soon after I started the fieldwork: Engaged in a job, time inevitably becomes scarce. Once out of work, time will be abundant. The following statement of an informant made me realise the importance of it: *When out of work, there is no spare time, no way to really have leisure. There is just a lot of time passing by. That's all there is.* Many unemployed spoke of their own lives as troubled by having *too* much time. They seemed to experience a time that heaped up, since it could not be put into any meaningful use.

In contrast to the theories of the *long durée* of Henri Bergson (2001), Gaston Bachelard (2000) argued that what makes us experience time are the interruptions of processes. For Bachelard the experience of lived time was fractured, interrupted, not single and continuous. He argues, in contrast to Bergson, that there is no underlying thread, that time is multiple and discrete. Retreating from the hassle of city-life to the quietness of a mountain area, all there is a continuous landscape, loneliness and no social interludes (Millerjord 2002). This experience is sometimes expressed in terms such as 'the time stops' or 'the time ceases to exist'. To some extent there is a parallel to be found in the situation of the unemployed. I will argue that it is primarily the social practices of work and consumption that make linear time into a social reality. Marginalised from these practices, one's life seems to be experienced as a continuous social landscape.

The personal career

To understand the aches and pains of unemployment today and how it is linked to discourses on and social practices of linear time, I will first dwell upon the concept of the personal career. What I want to propose is that a career more than anything else is an expression of success in self-improvement on all levels in life, that it has become the dominant mode of integrating the flexibility and complexity of post-industrial and globalised social life into individual lives.

To cumulate one's own career on a line is perhaps the most dominant way of expressing subjective autonomy today (Øian 2004: 186). It is a display of individual ambitions and achievements, rather than the conventions of one's social background and the anticipation that it will entail. The universality inherent in linear time makes the construction of pure individuality possible. The career has hence become to function as a productivist definition of the person given by the dominant social and moral requirement of accumulating different sorts of capital in a linear mode.

Unemployment can be looked upon as a time out from one's career. Some do in fact also perceive it in that way (Øian 2004: 179). It could be understood as one step out of the linear streaming of the time of the career. The time out could then be given meaning as a period of re-qualification for the labour market. In effect one could just re-enter the line of time at the same point where it was left, but stronger and more competent than before. However, and as I want to show below, social practices and discourses make this a difficult or even impossible project to most unemployed.

The individualisation and differentiation of late modernity makes individual lives less predestined and more unpredictable. Options and risks are manifold when it comes to paths and careers. In reality choices are made within a short time-perspective, and we often have to re-evaluate them. Still, things cannot be that fragmented and unstable in every respect. Most individuals engage in lifelong personal relations with sets of mutual obligations. Moreover an individual can change jobs or profession and the style of consumption, but the career can be perceived as unchanged in respect of its direction. In other words, and in accordance with the perspectives of Bachelard (2000), the personal career can be seen as a mode of constructing the *long durée* in order to integrate all the temporal interruptions of modern life courses. The personal career as such means much more than climbing the ladder within one and the same profession or company.

First of all a personal career can include social and symbolic aspects of the consumption of commodities, as lifestyles have become a more fundamental dimension of status and identity. In a modern consumer society, the integrating processes of a career should also include consumption of commodities and leisure. I'm not hinting at Bell's argument (Bell 1978) of an inherent contradiction in post-industrial capitalism between work and consumption, of which I disagree. On the systemic level consumption of commodities is a prerequisite for expansion of the capitalist economy. On the level of the individual, consumption has become the prime mode of engaging in the material world (Miller 1995). Moreover to some extent consumption has replaced work as a foundation of constructing identities and the meaning of social relations. The point here is that consumption in the context of life styles can be included in the personal linear career.

Secondly, a career can also include aspects of family life at home, but also relations to friends in the leisure spaces, just as much as interactions with employees, colleagues and market-partners. Thirdly, and most significant here, a career also means that through processes of the accumulation of symbolic, social and material capital, the individual makes a linear progress in time.

De-linearisation has been used as a term in order to characterise some of the changes caused by the post-industrialisation and globalisation of societies (see e.g. Appadurai 1990, Giddens 1991, Melucci 1995). Within these perspectives the concept of linear time tend to be taken for granted. The conception of the career, as I have explained above, does have some consequences for the perception of linear time as inherently uni-directional. In a typical industrial context uni-directionality of life was obtained by sticking to the same job or profession for the rest of one's life. Now uni-directionality rather refers to a constant self-improvement in terms of the accumulation of all kinds of individually held assets, within the frame of a personal career. The personal career may therefor maintain the linear perspective after all. But this is on a very individualised level, since linear time becomes meaningful in social practices of individual accumulation. Moreover, to the extent time units can be saved and invested, the career becomes symbolic and social capital in itself. Not to make time move by means of your own cumulative actions, to refrain from constantly producing a new present, means that the time runs away from you, which means a lack of individual autonomy and vigour. To be out of one's personal career is thus both a cognitive and moral problem.

All time becomes the same

A rather common feature among the unemployed informants was that they expressed disturbances over the experiences of time as socially *un*differentiated. Secondly, some were told by friends and families that there were no reasons to worry. *There will soon be a new job for you, just relax and enjoy it. You got all leisure now!* But no real leisure could be enjoyed, since leisure tends to be a meaningful category only in its opposition to work. Being unemployed is not only being out of work, but also being *out of leisure.*

Among the younger generation I could not find any unambiguous references to morality in these instances. In contrast to the generation of their parents, leisure was perceived as something that has its own rights, and not something that can be earned by foregoing hard work. Leisure is believed to contain activities just as substantial as work. In some respects the border between them becomes blurred, for instance when both are framed by the performance of lifestyles (Øian 2004). Consequently the experience of being out of leisure is not necessarily produced by moral categories but rather by cognitive categories.

Before unemployment one was always short of time. Even after work or school one constantly had to hurry up for appointments with friends or scheduled sport activities. Some were too engaged in a large spectre of activities and relations and did not get the sleep they needed. Thus, in some cases the initial period of unemployment period was experienced as a relief from all the constraints that school and job had put upon leisure activities. Freed from these kinds of obligations, some realised the dream of going to the clubs every night, while others became even more dedicated to the exercise of their sports or hobbies.

After a while frustrations emerged, nevertheless. One informant was a girl in the beginning of her twenties. She had been made redundant for a couple of years. She told me she did not worry that much when she had been laid off, thinking that another job would turn up, and that she had been comforting herself by the fact that she was entitled to an income that made her able to pay her bills. She did not identify herself much with the work, the work place or the colleagues. The job was more like a necessary prerequisite for realising herself in the spheres of leisure and consumption. She told me she had felt a mixture of relief and insecurity when she found herself out of work and had tried to look at it as an extraordinary holiday. In spite of the non-fatalistic attitude, she soon experienced problems she had not expected:

> I thought I could sleep a lot; do a lot of things there was no time for earlier – and the best of all – without any hurry. But it did not take long before I realised I just had far too much time. I could not enjoy all of it, because I could not find anything to fill it with. Time just heaped itself up. It did not really move. In the end it felt like nothing had its own time. The worst thing though, is that there is nothing much to look forward to. When at school I could always look forward to the end of the day. I always longed for the Fridays, and it felt like such a relief every time. Now I cannot have any real leisure, because there is nothing to have leisure from. All my time is spare time. But it is not proper spare time after all. In the long run, it does not feel OK at all. My time is not of any use. All time is the same.

All time becomes the same. Experiences of duration are tied to clock and calendar time only to the extent these chronometers do have some social relevance. Being unemployed there is no leisure, no pre-defined periods for rest and time-out. If time is not divided into work time and spare time, it tends to become indifferent whether the time is 4 p.m. or the calendar says Friday. The next hour or day do not feel different from the ones that just passed.

The same experience can of course occur while performing a boring job. There is one important difference, though: In spite of the dull routine of many jobs, the time is moving forward, driven by the clock and the calendar. All one has to do is to hang on and wait until the clock points at the end of working day. But without work, the calibrations of the clock and the calendar do not have the same social relevance.

Although time might not be socially experienced as linear, linear time is nevertheless retained as cognitive and ideological category. One of the worst things is the feeling that time is just passing by. The chronometers tend to be experienced as an abundant and empty form that can never be filled in meaningful ways. Completing a task on time or even before deadline is, for instance, not as meaningful a thing as it used to be. One does not really run out of time anymore. Rather it is time running away. Time is just passing by, but it cannot be put to any use and just disappears into a future, while oneself is left behind. The present becomes *continuous*, as Helga Nowotny (1994) has called it, and not something that is negated into the next minute, hour or day.

Linear time as social reality

Ontological ideas and categories do not have any social reality without being part of a social practice (Kapferer 1988: 80). The same is true about linear time to the extent it exists as ontology both on the level of cognition and ideology. What turns linear time into a social reality and not only a category or an idea, is first of all the social practices of work and consumption (Johansen 1984). Because these practices essentially involve money, linear time cannot be understood without taking into consideration how money implies universality and individualisation (Simmel 1978). To make it simple, the significance of money in this respect can be found in its contrast to an object whose value derives solely from its use. In a plough it is inscribed to plough, while in money almost any future possibility is inscribed (Johansen 1984: 133). In other words, it seems to be the dialectic relation between the chronometer and money that makes it meaningful to think of the future as a virgin field to be realised by individuals.

The relation between time and money occurs in processes where the orientation towards accumulation is dominant and provides the perception of a constant continuity between the present and the future. The forthcoming always means that a new present is produced. One is always directed into the future, always motivating one's actions by the expression "because of" (Johansen 1984: 102) since the present is relegated to a context for realising the future. In this respect linear time is a commodified time; time becomes money, and money becomes time. This is, however, meant both in a literal and metaphorical sense. Literally, as time is a scarce resource in capitalist production and accumulation. Metaphorically, as time can be saved and invested like a capital comparable to money. To be able to experience freedom from chronological time, one has to, by means of hard work and planning, save and invest time in order to earn time outs in terms of leisure.

Being unemployed hence means an everyday life where the most obvious practical references to time as linear are absent. Firstly, being unemployed just does not mean that one is out of work, but also *out of leisure* (Øian 2004). Leisure is a produced time and not something one just disposes of. It is not a free time in the true sense of the word, but it must be earned. Without a defined work-time, one cannot have a defined free time, neither in social nor moral terms. Secondly, without the daily engagement in work an important social structuring of motivations and directionality is lost (Bourdieu 2000, Øian 2004). I'm not hinting at the drive for the accumulation of capital only. The distinction between work hours and leisure as spare time is just as important. This distinction has surely a social and moral dimension, but also a cognitive one to the extent it creates self-evidence in the everyday (Øian 2004). Being employed, then, means an everyday life where there are no obvious practical references to time as linear (Bourdieu 2000: 221). What one does today or at this hour has no self-evident meaning as a realisation of tomorrow or the next hour.

The unemployed are subsequently trapped in the experience of a non-moving, empty time. This well-known common feature of unemployment figures in several studies and reports (e.g. Jahoda 1978, Pahl 1984, Seabrook 1982). As Pierre Bourdieu (2000: 223) pointed out, the unemployed are left with an abundant, but empty time only, because when time is not scarce, there is no time to invest.

The experience of time as duration

Another well known feature of unemployment is that everyday life appears meaningless. I often heard complaints about how one felt pacified and apathetic. Activities that used to be meaningful and joyful become dull and pointless. Traditionally this is interpreted as an effect of a general moral decay caused by the absence of the disciplining outcome of work. However, I think it can be interpreted better within the frames of how linear time, and the institutions that make it into a social reality, dominates our understanding of our self and the relation we have to our surroundings.

As already mentioned, Bachelard (2000) argued that the experience of time is not rooted in time as an a priori form or substance. Time is rather made up of our experience of interruptions and fractions of both natural and social processes. He maintained that time is not an integrated, one-dimensional continuity, but rather something produced by the experiences of discontinuity of various processes. Long durée, the lived time of Henri Bergson (2001), is therefore more than anything a social construct. How humans perceive the durations of processes is not identifiable to these processes themselves. Time as duration and continuity can be constructed in multiple ways.

Linear time, and the mode by which it is closely tied to the practices of work and consumption, make it possible to imagine that time can be saved or even produced. At the same time the present tends to be reduced to a context for realising the future. Each process of action that ends with an achieved goal means that time is brought further on; time has moved into the future and created a new present. This means that any action can be meaningful as an actualisation of the future, being the next hour, tomorrow, next month or year, and that actions are most of all accumulative. As such, any action can in itself represent the distinction between before and after (Gell 1992: 157). Before and after can in principle be about seconds or even a second split into thousand units, like in sports.

In a modern context of individualisation the concept of linear time should also include the conception of life as a cumulated stream into the future. To a certain extent this conception is closely tied to the linear career, as unemployment deprives one of the cultural and social means of saving and producing time in order to keep up with the chronological stream of time. The connection between the present and the future is not experienced as a stream of cumulative actions on a line. What is done now does not produce tomorrow. The time of the present is aimless and useless; actions of the present are not seen as vehicle of realising a future. One easily gets the feeling that time does not move, every day is the same. When tomorrow does not represent different experiences from today or yesterday, planning becomes futile. Any action, any motive get their meanings in the context of the present.

Since the relation between the present and the future becomes discontinuous, interactions and activities do not have any proper external excuse or motivation. Within practices of work, actions do have meaning in relation to an absent object that is to be realised in the future. Being unemployed, there is no object to be achieved as such. In effect one can neither experience time as brought forward on a line. There is no self-evident movement in time. Everything has to be initiated from the start. Time spent on whatever it is cannot be made meaningful in relation to the future. Duration is then rather experienced in relation to actions and events in themselves. To do something is not an instrument for accomplishing an object, but more like an end in itself.

Time as "good" and "bad"

Being deprived of one's cultural and social means of saving and producing time implies a lot of stress and frustration. Firstly, there are no predefined periods for relaxation or joy. Secondly, when the social relevance of the clock and the calendar diminishes, staying up all night and sleeping during daytime is a common effect. The latter implies that one tends to de-synchronise oneself with friends and others.

Studying the sociality of the urban everyday life the Swedish anthropologist Ulf Hannerz (1981: 102ff) characterised it by the term *traffic relationships*. On our way to and from work we interact with many persons who we do not need to have any personal knowledge of. In contrast to our families, friends and to some extent our colleagues, these relations require a minimum of engagement. Rather the avoidance of personal and emotional involvement is expected (Simmel 1971).

Nevertheless, these kinds of relations can be experienced as important, something which seems to become clear when it disappears. Several unemployed emphasised that they missed the occasional exchange of news with the shopkeeper or even joining the crowd on the bus stop in the early morning. Virilio (1997) and Schultz (1967) have, from different perspectives, theorised on how identification with *consociates* is fundamental for one's identity on an existential level. In these cases the marginalisation from the anonymous masses of the sociality in everyday life seems to be experienced as a loss. To remain in one's suburban flat while most others go to work in the morning and stay there until they return in the evening, can engender a feeling of being left behind. Some called it even an immense feeling of loneliness.

In addition, nobody needs the time of the unemployed. Others are away at school or work. In addition more and more services can be bought in the commodity market. The system of social exchange based on giving time is thus replaced by commodified and individualised time (Bourdieu 2000, Øian 2004). Although the unemployed have got plenty of time there are few opportunities of giving it to someone else.

Research has shown that the unemployed tend to withdraw from the public sphere and stick to their home and family (see e.g. Jahoda 1978, Seabrook 1980, Pahl 1994, Bech-Jørgensen 1993). In this manner one avoids the confrontation with the loss of the consociates, but also the feeling of being of lesser value, morally, than others. When time is empty and redundant, sleeping until the afternoon – as many start doing – makes sense in more than one way. It is not necessarily an automatic effect of idleness. It can also be interpreted as a rather consciously chosen mode of minimising frustrations caused by the marginalisation from the crowd.

In Norway leisure is still a time for retreating to one's private sphere (Gullestad 1992), to the home or the cabin out in the countryside. To some extent this kind of leisure is a retreat or time out from the scarce time of work. To have leisure or spare time in Norwegian is often termed *good* time ('god tid'), while being in a hurry is *bad* time ('dårlig tid'). When time is abundant, time is neither good nor bad in this sense of this word. Like one informant pointed out to me: *I do not have any good time, my time is just bad because I cannot do anything with it and I do not share it with anyone.* In some respect the employee's retreat to the home can resemble the retreat of the unemployed. The difference is that the former seeks to minimise frustrations, while the latter seeks to maximise pleasure.

Suspended time

Most unemployed people went to a so-called job centre on a daily basis. In fact the attendance at the centre was in practice compulsory, since the dole would depend on it. The centre was designed for multiple purposes. First of all, the unemployed were to be trained in searching and applying for jobs. There were classes in "How to set up a CV". The attendants were also guided in how to use a computer, copy and fax machines and telephones, which were available free of charge. Secondly, the staff at the centre made efforts in moulding the self-perception of the unemployed, making them realise their potentials and clarify their competencies and qualifications for the labour market. Thirdly, one clear aim of the centre was to impede laziness and apathy, which is taken to be the unavoidable effect of unemployment.

To many the centre was also a convenient place to socialise and gain confidence among equals. It is an important experience to know that one is not the only one. The atmosphere was a mixture of relief, intensity and optimism. Partly this was caused by the way the staff put emphasis upon self-responsibility for getting a new job. It was stated that they had been fooled by a myth, a myth claiming that there were not sufficient jobs for everyone. There are a lot of tasks around in society to be dealt with. It is all up to the unemployed to go out and present themselves ready to do the job. It was claimed that otherwise the tasks would not be realised and the jobs would remain invisible.

This individualising perspective upon unemployment was held within a neo-liberalist discourse. The attendants were to learn to sell their skills, competence and abilities, but also their personalities, on the market. Not everyone was familiar with writing in a formal style and took great pains in setting up a CV. Nevertheless, many were shining with pride and optimism when the CV was finished. Some were more reluctant and one attendant raised some critical comments, arguing that when every unemployed learned the techniques of selling oneself on the labour market, the competition would be just moved one level higher.

The job centre also tried to work upon the identity of the attendants. For instance it was held to be of outmost importance to substitute the term 'unemployed' with the term 'job seeker'. Being a job seeker is in itself a full time job more than any other, and it is a job without any spare time, as it was frequently asserted. *You cannot lie down and rest before the job is found. Something is going on in your head and body constantly,* one member of the staff claimed before a group. These kinds of statements were made several of the attendants disagree and complain. *This is not human*, a female informant in her mid-twenties said in an interview I recorded:

> Being without a job is made into something like a state of emergency. It is not my fault that I was made redundant, and so they tell me here. Still I feel I must be punished. I have to come here every day, or they don't pay me the dole, even though I'm not an employee here. And they tell me to continue with my efforts when I come home in the evening, as if I were not entitled to enjoy my leisure.

The time of the unemployed is seen as a time without any value in itself. Unemployed time is nothing more than an instrument for achieving the one and only goal. Still it is a non-generative time, since it is not found within the linear career proper. It is suspended time.

Conclusion

The concept of linear time can be divided into two aspects that are mutually related – but at the same time partly contradicting:

On the one hand there is a chronology of the clock and calendar, which is best seen as a technical device in organising tempo and timing of social practices of both sacred rituals and mundane production (Adam 1995). On the other hand there is also a dimension of linear time that refers to the cumulating of actions and events. The present is understood as a cumulated result of past actions and events. Likewise, the future has to be produced by cumulating actions and events. The former measures movement, changes and duration in quantifiable units. It is, as Barbara Adam (1995) emphasises, a *non-generative* time. It is an empty form that has to be filled with a content – in order to take advantage of it and transform it into a resource. The latter represents a social, qualitative and generative dimension of time. It does so together with other processes that create movements and change, like the biological, physiological, psychological etc.

Still, for the cumulating process to become a cognitive and ideological category, it is dependent upon the chronology that inheres in the metric measures of time: The clock and the calendar make it possible to envision life as a line from birth to death. Nevertheless it is not the clock or the calendar that has turned us into who we are, and it is not the chronological measure of time that is going to produce our future. Because the clock is non-generative in itself, it is human action that brings time forward. There is no chronological future; the future has to be made, as Richard Gault (1995: 156) has put it. It is rather the cumulating actions and events that create our lives to be envisioned on a line. Every process of action that ends with an achieved goal means that time is brought further on. Time has moved into the future and created a new present.

The way we experience time in modern society can thus be characterised by ambivalence. On the one hand chronological time imposes an objectified time that has its own life independent of human action. The time of clock and calendar moves on by its own force. In contrast to other times it is finite, it runs out and away from us (Adam 1995: 52). In other words, linear time is something we have to produce ourselves, by transforming the abstract chronometry into a resource. Otherwise it will make us feel left behind. In this manner linear time and clock time can be seen as an enemy – a kind of alienating force. On the other hand linear time can represent a resource just as much – something that can be saved and invested. The problem of the unemployed is that they are deprived of the social and cultural means of generating time, saving and investing it. That is why linear time can be felt as an enemy even more than before becoming unemployed.

Works Cited

Adam, Barbara (1995). *Timewatch: The social analysis of time*. Cambridge: Polity Press.
Appadurai, Arjun (1990). "Disjuncture and difference in the global cultural economy". *Theory, Culture & Society* 7.2/3, 295-310.
Bech-Jørgensen, Birte (1993). *Når hver dag bliver hverdag*. København: Akademisk Forlag.
Bergson, Henri (2001). *Time and free will: An essay on the immediate data of consciousness*. New York: Dover.
Bourdieu, Pierre (2000). *Pascalian meditations*. Oxford: Polity Press.
Gault, Richard (1995). "In and Out of Time". *Environmental Values* 3.4, 149-166.
Giddens, Anthony (1991). *Modernity and self-identity*. Cambridge: Polity Press.
Gullestad, Marianne (1992). *The art of social relations: Essays on culture, social action and everyday life in modern Norway*. Oslo: Scandinavian UP.
Hannerz, Ulf (1981). *Exploring the city: Inquiries towards an urban anthropology*. New York: Columbia UP.
Jahoda, Marie, Paul F. Lazarsfeld & Hand Zeisel (1978). *Marienthal: The sociography of an unemployed community*. London: Tavistock.
Johansen, Anders (1984). *Tid är makt, tid är pengar*. Malmö: Röda Bokförlaget.
Melucci, Alberto (1996). "Youth, time and social movements". *Young* 4.2, 3-14.
Miller, Daniel (1995). "Consumption studies as the transformation of anthropology". *Acknowledgning consumption: A review of new studies*. Ed. Daniel Miller. London: Routledge, 264-88.
Millerjord, Terje (2002). *Mot en teori om tid: Basert på Gaston Bachelard, Pierre Bourdieu og et feltarbeid blant fjellturister i Nord-Sverige*. Master Thesis, Department of Anthropology, University of Tromsø, Norway.
Nowotny, Helga (1994). *Time: modern and postmodern experience*. Cambridge: Polity Press.
Øian, Hogne (2004). "Time out and drop out. On the relation between linear time and individualism". *Time and society* 13.2/3, 173-195.
Pahl, Raymond E. (1988). „Introduction: Work in context". *On work: Historical, comparative and theoretical approaches*. Ed. Raymond E. Pahl. Oxford: Basil Blackwell.
Schutz, Alfred (1967). *The phenomenology of the social world*. New York: Northwestern UP.
Seabrook, Jeremy (1982). *Unemployment*. London: Quartet Books.
Virilio, Paul (1997). *Open Sky*. London: Verso.

Steven L. Arxer, John W. Murphy & Linda Liska Belgrave

Globalization and the Politics of Time

Introduction

A persistent but subtle mode of social control is now operating in the world. The central thesis of this paper is that this type of control is consistent with the view of time that has become dominant – namely, Newtonian time (Price 1996). For accompanying this perspective on temporality is social imagery that may lead to repression and a truncated future. Newtonian time, in other words, can establish an absolute framework that can be used to reinforce norms that are not subject to critique. Given this a-historical context, these rules can be described as part of human nature or associated with functional social imperatives. In any case, norms are provided with a universal status that stifles criticism and reifies order.

What is crucial is that this form of control is made to appear apolitical, since, as Foucault (1977) suggests, physical manipulation and overt coercion are inconsistent with the modern era and the rule of law characteristic of democratic societies. Nonetheless, without violating the implied pluralism, social order has to be enforced. In this regard, a method of control is used that is thought to be unfettered by human foibles, values, and biased political interests, and thus social constraint appears to be divorced from any particular perspectives or personal desires. If norms are understood to be severed categorically from social or cultural considerations, in other words, their political character is easily obscured and assimilation to developmental or other norms is considered to be rational and even a virtue.

Through the principles of Newtonian time, social order is ensured because a developmental or historical scheme is sustained by an absolute temporal model rather than personal preferences. What is important at this juncture is that this pursuit of universals has fostered an extremely successful way to restrict change. Simply put, a specific mode of dualism is invoked that makes particular images of social existence appear absolute and effectively undercuts ideas and practices that are antagonistic to these universals. What makes this maneuver so powerful is that control is made to appear rational and even necessary. And because force appears to be absent during this process, this tactic is referred to as symbolic control by some writers and as "symbolic violence" by others (Thompson 1984: 42-72).

Symbolic Control and Time

Nowadays persons are told that they should strive to adapt to historical necessity, thereby guaranteeing their social acceptance and success. At the heart of this process of normalization, however, is that conformity is achieved through the use of language. In this case, language should be understood broadly to include speech, images, and logic. What occurs is that people become enamored of particular symbols. According to Bourdieu (1990: 84-85), "symbolic [control] [...] is [control] exercised [...] in formal terms". In this context, "formal" means that the "force of the universal" is intimately tied to the "force of the official". The key consequence is that through the invocation of specific universals, an allegedly neutral method is available to both justify certain ideals ("the force of the universal") and to enforce them ("the force of the official").

To use Guattari's (1984: 168) words, "dominant signifiers" are employed to subvert critique and alternative ways of thinking, speaking, and acting. Moreover, because this abstract conception of knowledge is believed to be superior to all others, the accompanying version of social reality is easily enforceable. With such an exalted status, this monolithic reality quickly garners support from people who simply want to act rationally and receive social confirmation.

Now the question is: How are particular symbols – such as an image of time – able to control reality? Stated simply, select images are able to monopolize reality through a fatuous theoretical position; that is, dualism – based on the distinction between fact and value – is considered to be legitimate. What is important to note is that as a result of this philosophical maneuver a hierarchy is created between knowledge bases, whereby those that seem unburdened by interpretation and other human shortcomings are believed to be objective and mostly credible.

To achieve this objectivity, many societies have opted typically to use structural or mechanical images of time when describing progress (Adam 1990). In particular, Schmidtke (1999: 169) contends that "[t]he Newtonian model of time is the currently predominating conceptualization within [...] all the social sciences". According to Hendricks (2001), to adopt a Newtonian model of time means to have an exact chronological image of the universe. By this he means that time is understood to be primarily linear and operate like a clock. In this sense, Lacan (1977: 77, 98) explains that when time is viewed as a clock, temporality is "mathematized". Consequently, time is viewed as a measure of distance, and temporality represents the side by side placement of an infinite number of moments.

Furthermore, these moments resemble points on a temporal line. And as is commonly assumed, there is an exact present that resides between the past and the future. This is what William James (1950: 608-609) referred to as the "knife edge present," for as movement occurs along this timeline, present after present emerges and disappears. Each present is self-contained, and because each moment is discrete, no moment should be mistaken for any other. As a result, each present is considered to be universal, since only one present can exist at any instant.

What becomes available to people is a foundation of reality that is purportedly apolitical and untainted by other human eccentricities. This temporal universal, in short, represents a baseline from which standards can be derived that all rational or normal people should mirror. Rather than arbitrary, these ideals are given seigniorial status. Within a mechanized image of time, this reification of reality occurs in a particular way. First, because time is linear, events can be organized in a simple causal sequence. And second, given that each present is universal, an autonomous temporal basis exists to justify order, along with any changes. Therefore, society is preserved and integrated when people arrive at a similar present, since their actions are synchronized.

An important consequence of this dualism is that specific meanings are permitted to overshadow all others. A central point, therefore, is that other perspectives must be rejected, for it makes no sense for people to adopt ideas that are out of step with the times. And because an asymmetry is created between knowledge bases, a specific version of history is legitimized and opposition effectively quelled.

The idea is that these universals do not represent a political viewpoint; for instance, Newtonian temporality is envisioned to be objective and have unrestricted validity. The end product of this process is what Marcuse (1964) calls a "one- dimensional" world. Within this realm a monolithic reality reigns and all challenges are undermined through official and, thus, socially acceptable means.

Symbolic Control in Practice

Up to now the logic of symbolic control has been outlined. This next section is dedicated to exploring how symbolic control may be manifested with respect to understanding societal and historical trends. In line with this discussion, a mechanical and mathematical view of time is usually employed to justify certain social changes, because of the accompanying autonomous timeline. As Hawking (1988: 18) explains,

> Newton believed in absolute time. That is, [he] believed that one could unambiguously measure the interval of time between two events, and that this time would be the same for whoever measured it, provided they used a good clock.

Consequently, for Newton, time serves as an unquestioned frame of reference for charting the movement of objects and people.

Hence any predictions related to the present or future, for example, are understood to be inherently valid, since this information is assumed to be based on an objective trajectory. In the end, proposing a naturalized portrayal of time provides the theoretical grounds for literal descriptions of the options available to individuals and groups. The central point is that change is natural and that inherent correctives are present to assure that development does not become disruptive.

At this time, much of the world is suffering from the effects of globalization, and at the heart of this trend is the philosophy of neo-liberalism. What persons are told is that this process is inevitable, even beneficial. Furthermore, in some conservative circles this process is touted as irresistible and irreversible (Johnson 2004: 260). This claim is predicated on the idea that globalization represents the culmination of both short and long-term cycles, which result in increased economic efficiency and prosperity. Additionally, given this description, only those societies that are irrational and unwilling to join the modern world would dare to resist this trend.

The logic of globalization, as an important historical trend, is thus made apparent to all rational and informed persons. This process of modernization can be ignored only by those who are foolhardy and uninformed about the realities of economic expansion. According to this scenario, history can be turned backward, but at a high and unacceptable cost to human development.

In many ways, this perspective on development is reminiscent of Hegel's view of history. The idea is that, simply put, history delivers only rational and beneficial ends. And at the root of this portrayal of change is a Newtonian image of time. That is, history reflects a timeline that is autonomous and capable of transcending contingencies and disputes, so that the future is clear and enticing. Globalization, in this sense, is entirely legitimate and represents the best that humanity has to offer.

The market is the centerpiece of neo-liberal economics (Cox 1999). And as a product of history, the market is autonomous, and thus is immune to biases and political intrigues. A society that is regulated by such a mechanism, accordingly, is progressive and in the forefront of economic development. The introduction of markets into every sphere of social life, from hospitals to schools, will culminate in society being regulated rationally and profitably. How can anyone who is sane question these outcomes? At least this is the message that is broadcast to critics of globalization.

In this sense, these persons are equated with Luddites. They are thus retrograde and fighting a useless battle against inevitable forces. They have no chance of success, but can certainly cause a society to miss certain key economic opportunities. These rebels are simply dismissed as unable to adjust properly to the times and are harming their societies. They are, possibly, idealists who are terribly out of step with history.

An Alternative Image of Time

The end result of this symbolic control is that people are disembodied. That is, people's actual experiences and identities become obscured as they try to emulate an idealized image of life. As noted by several writers, this unfortunate state of affairs can be avoided only by subverting dualism (Bordo 1990).

In order to implement this theoretical gambit, the a-historical referent used to distinguish fundamental from accidental traits must be revealed to be illusory. The problem with dualists, however, is that they ignore that all knowledge is mediated by *praxis*. But phenomena such as concepts, facts, and laws do not emerge untainted by perception. Indeed, perception does not represent a passive encounter with reality. As Bergson says, in addition to interpretation, perception engages and transforms whatever is known. Everything from historical portrayals to personal and group commitments to a particular version of time and society shapes the nature of facts. Before reality makes sense, this normative base must be conceptualized and placed in a framework, and this activity is fundamentally an interpretive process.

As a result of reality having this interpretive base, Lacan (1977: 75, 98) questions the legitimacy of linear, clock time. He argues that when time is described as linear, temporalization is misrepresented. Borrowing from Bergson, Deleuze also says that the passage of time fails to mirror a clock. Similar to Bergson, Deleuze (1988: 51-72) argues that time is *durée*. Rather than a measure of distance and space, the movement of time consists of shifts in conscious attention and focus.

Accordingly, the past is not something that disappears behind a time horizon, but is a present that has lost some of its importance or intensity. Instead of a line, time represents a montage or collage of significant periods. In this sense, moments are not discrete entities, but rather exist in what might be called an "expansive present". Consistent with many existentialist writers, Luhmann (1979: 10-17) points out that moments in time are not disconnected; time exists as the past-present, present-present, and future-present. Every moment of time is underpinned, in other words, by how temporal movement is understood. Hence time does not push events forward; rather, time is a delimited expanse of action.

Lacan (1977: 44, 48, 75) also believes that time is experiential. He notes that temporalization represents the constant renewal of the present. According to Lacan, there is a difference between the "*temps pour comprendre*" and the "*moment de conclure*". What he means by this distinction is that temporal coordination depends on individuals agreeing about the nature of time, rather than simply arriving at a similar or identical present.

When time is imagined to be *durée*, the present is constructed and maintained in the midst of other competing possibilities. In de Man's (1972: 239-67) words, time is the "*représentation du présent*". Consequently, there is no *a priori* present; the present has to be brought into existence. For example, temporal change depends on people having a common view of temporality. Temporality, therefore, is foremost the experience of time. As Deleuze (1972: 115) points out, "[t]ime is precisely the transversal of all possible spaces, including the space of time". He goes on to add that time represents the "ultimate interpreter, the ultimate act of interpretation".

The main issue at this juncture is that because temporality is tied to the consciousness of people, the ability to understand their behavior depends on grasping their respective conceptions of time. Accordingly, as Husserl (1964: 28) argues, the passage of time must be approached as changes in the "lived experiences of time," instead of the movement of an eternal clock.

With this new rendition of time, the foundation of symbolic control is undermined, and the ability of particular knowledge bases to inferiorize others is challenged. Because nothing, not even Newton's absolute time, escapes the influence of interpretation, all portrayals of the future, for example, rest on the same ontological plane. That is, none are inherently superior to others and able to demand automatic recognition. As Schutz (1964: 135-58) claims, from a myriad of possible descriptions, any one of these can become "paramount" through human volition.

Following this reasoning, Lyotard (1984: 9-11) suggests that time exists as a "language game". What this means is that similar to a game, temporality is a matter of convention. What was once thought to be natural is now historical in origin. This rendition of time is liberating, since people are free to define themselves as they see fit; however, they may also choose to accept or be cajoled into adopting the prescriptions of others. In any case, what is important is that the so-called normal expectations of social existence can be rejected without people being inferiorized. Adopting a traditionally aberrant conception of the future, something considered utopian, does not necessarily signal an irrational position. Horkheimer and his Frankfurt colleagues contend that this way of examining social reality is mature, because legitimacy does not come from an external source of authority but from the critical reflection of the populace (Jay 1973: 119-20).

Embodied Development.

What proponents of autonomous time have done is try to dehistorize norms and establish universal developmental guidelines. The reductionism that has resulted is anathema to the diversity of possibilities that characterizes the future (Massey 2005). But almost by definition, the future defies control.

On the other hand, Lyotard (1984: 61) argues that all norms, and thus history, are "locally determined". Rather than uniform and continuous, as linear time would suggest, shifts in language use make reality discontinuous. Reality is not smooth, but rather is more like a patchwork of competing views. Accordingly, norms are situated in particular meaning-patches, and the norms derived from Newtonian time are no exception. These norms are not automatically generalizable simply because they are believed to be autonomous, but rather represent one among many ways of depicting the past, present, or future.

But when time is experiential and intimately tied to consciousness, development does not represent an unrelenting line. For rather than a measure of distance and space, the movement of time consists of shifts of consciousness. According to Merleau-Ponty (1962: 412), "[t]ime is not a real process, not an actual succession which I record. It arises from *my* relation to things". Time is not comprised of discrete moments that are disconnected, but rather represents modes of consciousness wherein the past, present, and future interpenetrate one another. The important consequence of this change in thinking for Heidegger (1962: 392-93) is that when tied together, these moments function as *enabling possibilities*. What he means is that rather than predestined and ever-diminishing, the future is given validity and meaning by the present.

This rendition of time offers a new way to conceptualize social change. Since human *praxis* extends to the core of time, moments do not merely pass away but are ordered with respect to their relevance. The future, therefore, is thematic rather than a point on an autonomous and unrelenting timeline. No period of time is more valuable than any other, and thus none offers any more or fewer opportunities. What is signaled by time is the ability of persons to make themselves in any number of ways. Time, in this sense, never closes horizons of possibility. Globalization is thus a story with many possible endings rather than a single trajectory.

Social change, accordingly, is not a product of a universal temporal line of development. Something like globalization, instead, constitutes a "signifying system" and the outgrowth of how time is experienced (Williams 1989: 13). This change in viewing temporality alters significantly how the past and future are organized. The guidelines that people follow are still meaningful but no longer exist *sui generis*. Within the perspective of experiential time, globalization represents merely another transitory means of expressing human action. In this respect, people do not passively attain meaning by following or adapting to a specific historical trajectory, but rather through interpretation particular modes of behavior and conduct are given significance and established as meaningful, at least temporarily.

Consistent with this shift in understanding time, there are nowadays various movements "from below" that are trying to change the course of history. These groups believe that they can redirect history, instead of merely adapting to economic or other inevitabilities. By taking this initiative, they have become the new subjects of history, and do not view themselves as pawns of an abstract historical *telos*. They want to make a better world that, at this moment, seems to be impossible.

The term used in Latin America, for example, is that these movements are "prophetic". The point is that they are not a product of history, but instead can alter this process. They erupt into chronos, or time, and make the future, instead of simply waiting for their destinies to arrive.

They have the ability to manufacture a new history; they are not controlled by the future but project entirely new possibilities. History is thus decidable, instead of a-historical and undecidable.

Conclusion

The significance of drawing attention to symbolic control is twofold. First, social control has been understood traditionally as either overt or covert. In both instances, a conflict of viewpoints is implied whereby the exercise of power determines the winner. At the heart of symbolic control, however, is the concealment of power all together. In the modern era, the use of force to achieve an end is considered improper and passé. A more sophisticated means to control the ideas that circulate in society is to convince persons that a particular type of knowledge is inherently correct while others are not (Habermas 1973: 263-76). Central to symbolic control is the imposition of an allegedly neutral portrayal of proper expectations.

And second, raising the notion of symbolic control exposes the dualistic nature of Newtonian time whereby (1) knowledge bases are asymmetrically organized, (2) rationality is cleanly differentiated from irrationality, and (3) an inviolable reality is erected that discredits automatically all oppositional norms.

Undermining the autonomy of time, therefore, provides globalization with a new image. The current phase of neo-liberal economics is neither inevitable nor universal, but representative of certain commitments related to how persons should view themselves, other persons, and social development. There is nothing magical, in short, about this activity. Furthermore, the changes that many oppositional groups from below desire are not necessarily unreachable or unworkable. The "post-capitalist" logic that is often discussed is not automatically wildly utopian.

The point is that when time is understood to be experiential, persons can remake the world, including their destinies. In the absence of an absolute temporal referent, stated differently, new norms can be constituted as historical and possible. This is not to say, however, that agents of change – those, for example, who want to struggle against inhumanity of globalization – will not encounter resistance. Of course, those who benefit from the current course of development will likely try to discourage change by any means possible.

What is important, however, is that any obstacles that are erected cannot be given the usual metaphysical status. That is, they cannot be treated legitimately as natural or social imperatives, thereby subverting any new future that might be envisioned. When norms are rooted in the experience of time, they are understood to represent one among many perspectives, and thus are always subject to redeployment.

In terms of this theoretical background, what many protesters have decided is to take their future into their own hands. And a more experiential version of time reinforces the idea that the future can be reinterpreted and redirected without the stigma that critics and activists are behaving irrationally and irresponsibly when they challenge current global trends. They may be violating the expectations of those who hold power at this moment, but their vision of the future is not without merit. Based on experience, the future can be reinterpreted and take any form or direction. Hence instead of dictating the future, time allows various destinies to be produced.

Works Cited

Adam, Barbara (1990). *Time and Social Theory*. Philadelphia: Temple UP.

Bordo, Susan R. (1990). "Postmodernism and Gender-Skepticism". *Feminism/Postmodernism*. Ed. L. J. Nicholson. London: Routledge, Chapman & Hall, 133-156.

Bourdieu, Pierre (1990). *In Other Words*. Stanford: Stanford UP.

Cox, Harvey (1999). "The Market as God". *Atlantic Monthly* 283.3, 18-23.

de Man, Paul (1972). "Literary History and Literary Modernity". *In Search of Literary Theory*. Ed. M. W. Bloomfield. Ithaca, New York: Cornell UP, 239-67.

Deleuze, Gilles (1972). *Proust and Signs*. New York: George Braziller.

Deleuze, Gilles (1988). *Bergsonism*. New York: Zone Books.

Foucault, Michel (1977). *Discipline and Punish*. New York: Pantheon.

Guattari, Felix (1984). *Molecular Revolutions*. Middlesex: Penguin.

Habermas, Jürgen (1973). *Theory and Practice*. Boston: Beacon Press.

Hawking, Stephen (1988). *A Brief History of Time: From the Big Bang to Black Holes*. New York: Bantam Books.
Heidegger, Martin (1962). *Being and Time*. New York: Harper and Row.
Hendricks, Jon (2001). "It's About Time". *Aging and the Meaning of Time*. Ed. S. H. McFadden & R. C. Atchley. New York: Springer, 21-50.
Husserl, Edmund (1964). *The Phenomenology of Internal Time Consciousness*. Bloomington: Indiana UP.
James, William (1950). *The Principles of Psychology*. New York: Dover.
Jay, Martin (1973). *The Dialectical Imagination*. Boston: Little, Brown.
Johnson, Chalmers (2004). *The Sorrows of Empire*. New York: Henry Holt.
Lacan, Jacques (1977). *Ecrits*. New York: Norton.
Luhmann, Niklas (1979). *Trust and Power*. New York: Wiley.
Lyotard, Jean-François (1984). *The Postmodern Condition: A Report on Knowledge*. Minneapolis: University of Minnesota Press.
Marcuse, Herbert (1964). *One-Dimensional Man*. Boston: Beacon Press.
Massey, Doreen (2005). *For Space*. Thousand Oaks: Sage.
Merleau-Ponty, Maurice (1962). *Phenomenology of Perception*. London: Routledge & Kegan Paul.
Price, Huw (1996). *Time's Arrow and Archimedes' Point: New Directions for the Physics of Time*. NewYork: Oxford UP.
Schmidtke, Charles R (1999). "A Pulsational-Wave Model of Time for Theories and Therapies of Aging". *Journal of Aging and Identity* 4.3, 167-83.
Schutz, Alfred (1964). *Collected Papers*. Vol. 2. The Hague: Nijhoff.
Thompson, John B (1984). *Studies in the Theory of Ideology*. Berkeley: University of California Press.
Williams, Raymond (1981). *Culture*. London: Fontana.

agency and resistance

Agata Stopinska

Genders in Flux

Polish and Irish Velocities of Doing Gender through Control and Rebellion

> Individuals have many social identities that may be donned or shed, muted or made more salient, depending on the situation. One may be a friend, spouse, professional, citizen, and many other things to many different people – or to the same person at different times. But we are always women or men – unless we shift into another sex category. What that means it that our identificatory displays will provide an ever-available resource for doing gender under an infinitely diverse set of circumstances.
> [Candace West and Don H. Zimmerman]

Gender is – according to West and Zimmerman's hypothesis – not primarily an achieved status by means of psychological, social and cultural influences but rather a result of one's immanent activity of managing situations in which conduct is regulated by normative conceptions of attitudes and activities meant to be appropriate for one's sex category in a specific culture in a given time. Women and men are not merely gendered, they do gender themselves, often reflecting social expectations. Yet *being gendered* or rather *doing gender* is regulated by traditions and antitraditions shaped by societies, families, religions, educational institutions of a culture and it gets transmitted, altered or preserved in time and space. Assumptions about gender and its constitutive elements, that we take for granted, are a product of such cultural regulations and their validity underlies constant actualisations. As West and Zimmerman (2004: 150-168) show, in every single interaction and every single act one is always perceived primarily as a gendered human being. Once differences between females and males have been defined and accepted, they in turn become instruments of reinforcing these poles, reinforcing the meaningfulness of gendered spaces that they form through social interaction. Scrutinizing these practices, West et al. try to give a back seat to the socio-cultural influence on an individual and foreground individual agency arguing "that a person's gender is not simply an aspect of what one is, but, more fundamentally, it is something that one does, and does recurrently, in interaction with others" (West and Zimmermann 2004: 161). But which agencies dominate which spaces and why? Does feminist agency opposing a state's policies prove the rebellious disposition of the discriminated? What agency of the discriminated is needed for a shift in power distribution, a shift from "periphery" to the "centre", from "down" to "up" to show up? At what point can females and males, in the age of gender equality policies, relinquish the sedimentation of conventional gender roles and set foot in new spaces?

The concepts of femininities and masculinities result from spatial performances correlated with constant validation by the public, by other persons, as well as with the given grade of internalization of conventionalized gender roles. This spatial character of genders, in the sense of dominant and alternative spaces that both genders occupy in a society, determines the process of integrating certain roles. The formation of these roles depends on the desire of being accepted by a social group one intends to belong to and is based on a life-long, often unconscious, reproduction and projection of the conventionalised. Women are not feminine by default, they respect certain rules of conduct and display their femininity publicly, since particular forms of gendered behaviour are expected at this very time in this very society and culture they belong to. The same applies for men. Under particular historical conditions, social control exercised by the state or/and the Church is involved in the production of the desired symbolic gender imageries, and in the renegotiation of roles attributed to men and women. As a result femininities and masculinities become carefully constructed forms of gender, either almost instinctive mechanical displays of social expectations or strands of gender discourse framed, controlled by the state, mass media, Church, popular culture etc. However, the conventionalised appearance, behaviour, mental patterns and other gender-determined elements can – under certain time- and space-related conditions – be rejected and challenged. *Gender* as such must therefore be analysed as a maze of interconnections of time, space, and agents of different age, race, class, dispositions that construct it, that *do* it in everyday interactions and model social discourses around it, often controlled by institutional powers. Furthermore one has to keep in mind the fact that the process of gender formation has two main sources: the level of everyday practice and the level of the gender discourse set both in the mainstream culture and in its subcultures. In everyday practice individuals may act according to available legal ordinance and their actions are either defined by formal opportunities and limitations introduced by local law, authoritative powers (introducing particular regulations within social welfare, the shape of the labour market, childcare facilities etc.) or are prescribed by other institutions that act for instance in the role of moral authorities (such as the Church). On the discursive level, individuals are offered symbolic, normative, allegorical gender ideals, and spaces ascribed to them, today also by media and advertising. Understanding cultural meanings of *gender* therefore means fathoming the external agency, too, as e.g. the control by the state or legislature, and the mechanisms that rule these culturally transmitted ideals.

Trying to analyse the problem of an external (controlling) agency, we have to start from the assumption that the process of becoming feminine or masculine occurs in spaces of mediated discourses which may be produced in the interest of a narrow community or a wider global market. Becoming, being and performing femininity or masculinity means different things in different spatial and temporal dimensions involving manifold settings of nations, ages, ethnicities and religions. Such discourses around gender involve the construction of femininity and masculinity as polar opposites and essentialize these differences. Doing gender, which Judith Butler identifies in her theory of perfomativity as "the repeated stylization of the body, a set of repeated acts within a highly rigid regulatory frame that congeal over time to produce the appearance of substance, of a natural sort of being" (Butler 1999: 43)[1], is unavoidable and the state of *being sexed* and through this, *being gendered* is unavoidable. The sex-category membership determines arenas of power, influence, recourses, privileges etc. as well as interpersonal interactions in a society. Gender as a social construct is a dichotomous and binary set of characteristics, a system of meanings that governs interpersonal levels of communication, shapes the self-concept and generates various gender-types. Secondly, gender is a set of properties that ascribe socio-cultural roles, traits, types of behaviour to sex-categories and are normative for members of these sex-categories in a culture. Gender cues shape the behavioural patterns towards others, even if only little of this sex-categorization takes place in the individuals' awareness. Physical appearance of women and men does not have to be on par with performative concepts set in their temporal and spatial contexts.

[1] Gender is done in Butler's view not by an agent, by a subject, a "doer" who might pre-exist the act of doing gender but it is rather a doing without an agent. Therefore she insists on using the term *performativity* and not *performance*. For her the doers are only fictional, she maintains that "there is no gender identity behind the expressions of gender; that identity is performatively constituted by the very 'expressions' that are said to be its results" (Butler 1999: 33). According to Butler, identities are constructed through performace and gender is a regulated fiction which gets formed anew in new contexts, new spaces and at different times. It gets articulated by performing it in everyday acts that underlie social and cultural sanctions and regulations. Butler regards gender not as an essential constant but as a fluid variable that changes at different times and in different localities. Out of this reason gender is performative, it is not an attribute of an individual in the sense of who you are. Gender must be seen as a fluid category which is determined by the fact of being placed and spaced.

Gender is essential for interactions with others across spaces of habitus, it affects one's life as well as creates social expectations and mental structures. It is time and space that construct certain roles, manifold gender performances, conventions, rituals, forms of conduct, moralities, and also deconstruct them. These shifts and a constant cultural flux reformulate gender, give it a socially scripted dramatization of the idealization of *femininities/masculinities*. Particular locations, temporal conditions – as shown in the following exemplary analysis – allow certain gendered performances that different locations at different times would not permit.

The once emptied – because censored – space around gender issues in both Polish and Irish public discourse, has been gradually permeated by topics that before had no likeliness to appear in public. These spaces now seem to have become more visible, public discourses shape and contest new meanings, new discursive spaces are created, once "deadly secrets" become "an object of analysis and a target of intervention" (Foucault 1999: 517).

The postmodern moral spacing, as Zygmunt Bauman (2003) argues in his book on the liquid nature of relationships in the age of modernity, engenders a relativity of values and systems of ethical concepts. Moral responsibilities of agents also underlie transformations and the traditional discourse gives way to progressive postmodern discourse culminating in a range of contradictory attitudes towards gender relations. A more liberal, softer policy-making (as for instance in postsocialist Poland) brings about alternative meanings to gendered matters. The media encourage today's societies, and first of all the young generation, to be sexually attractive, active and desirable. One may have the impression that these societies are most concerned about the issues of relationships with the opposite sex, about the pleasures of having sex, looking attractive, being alluring in reciprocal contact. Conventional Christianity-based moral concepts of "classic" modernity become less meaningful in the era of the postmodern world of media, new openings and redefined categorizations. Those not conforming to the social expectations are "waste" (Bauman: 2004), that can be thrown away, since they are not real agents of their lives.

In this paper I would like first to look at concepts around femininities and masculinities in two different but still very similar cultural settings, and do this partly by looking at linguistic realizations of these concepts. Due to the fact that it have been women, who much more often than men became objects of public and private debates concerning their role identification, the mutual relation between genders comes to the fore by documenting women's ideals. Secondly, I intend to map changes in gender issues within the last thirty years both in Poland and Ireland, two peripheral, traditionally profoundly Catholic European countries, linked until recently by conservative agricultural backwardness and politically strongly influenced by external powers. The paper tries to analyze the cultural, socio-political contexts and sketch a rough contrastive study of these two settings. The societies of countries between Poland and Ireland have gone through a clear-cut sexual and moral revolution since the 1960s, which brought about what Rose calls "Enttraditionalisierung-, Ausdifferenzierungs-, Individualisierungsprozesse der Frauenrolle" (Rose 1997: 126), whereas the periphery – for political and other reasons – long years after the changes in the West, remained frozen in its gendered codes as in a time warp. Deliberate, politically or ethically motivated silencing of particular voices occurred on both edges of the European periphery. Western Europe over many years was framed by two peripheral areas of retrogressive conditions, albeit retrogressive in different ways. Poland and Ireland in a similar time experienced a radical "reconstruction of western societies in terms of occupational structures, secularisation and demography" (Tovey 2000: 188) as well as encountered "the rise of the modern women's movements and the permeation of feminist and gender-oriented discourses into all aspects of social life" (ibid.). In this process media played a crucial role as – after the abolition of censorship – they transmitted opposing codes, alternative modes of re-thinking reality and they were able to reach masses after the introduction of television. Both societies – due to the late achievement of free media, a long-term censorship from diverse sources and a kind of cultural imperialism – have been experiencing social change for barely or just over two decades. Nevertheless they have been going through enormous transfigurations: both originating in homogeneous, culturally closed and deeply Catholic societies and turning into open, pluralist and globalized cultures. The modernisation of Polish and Irish societies is gaining ground, and in spite of different political, historical, social and cultural conditions setting those two places apart, – in certain aspects – these changes show surprisingly many similarities.

"Mother Poland" Liberated?

The norms of gender-related concepts in Poland establish certain ideals which are linked to the culturally constitutive ascendancy of the middle class and the ideology of domesticity. By the end of the nineteenth century – after long struggles over conduct and respectability – certain prevailing respectable forms of white middle and upper-class femininities and masculinities had been formed. In socialist Poland these norms established before the second world war mingled with the official policies of the Communist regime, and despite the state's efforts to dominate and determine all areas of life, new values imposed upon the post-war Polish society could never be fully absorbed and, as result, formed a hybrid with the traditional ascendancy of the bourgeois. These new values connected to the Communist ideological, political, social and economic programme stressed the importance of introducing work possibilities for women.

The post-war economy that suffered from an unspeakable loss of population needed labour for purposes of rebuilding the country, its destroyed cities and of implementing plans of industrial reconstruction. Additionally, Marxist ideas of social equality fitted into a policy that encouraged women to take up positions of paid employment. Many women turned to wage labour since one person in a household was not able to maintain it and men lost their role of the breadwinner in the pre-war period. In the 1960s the state had to react to the baby-boom and introduced legal contraception, promoted – in spite of the Catholic Church's opposition – birth control and the "2 plus 3" model of a nuclear family. These changes resulted in creating and enforcing the image of an employed mother. The official model of desirable womanhood was – according to Marody et al. (2000: 157) – an employee and mother whereas the ideal of a good mother seemed to be more attractive and serve "real" self-fulfilment. This double female identification, shaped carefully by the state, in the state's policy for gender relations found no counterweight in terms of concepts of masculinity. It was implied that the man was the "head of the family", at least externally, and that women accepted with pleasure their primary roles of wife and mother. The vision of a woman's self-sacrifice, her devoting herself and all her agency, individual needs, ambitions etc. to the needs of her children, family and country dominated the post-war concept of motherhood. Heroic motherhood, the allegoric Polonia, "Mother Poland", as well as employed women were the key concepts around femininity under the Communist regime in Poland. Anything different was already a deviation from accepted standards, as for instance divorced or unmarried women, not to mention lesbian women. Consequently, state socialism is often imagined as a totalitarian matriarchy with little private sphere, in which women could exercise their symbolic power, whereas home and family under the same regime are symbols for a totalitarian patriarchy. Men were perceived as those who enjoy public agency, their positions of power, but at the same time, in practical tasks, as small boys, anti-agents who at home are not able to take on any household duties, and therefore desperately need female help.

Only after the major systemic upheaval in 1989, a third radically different element – after the pre-war morality of being chaste and modest, and the post-war socialist emancipatory strategies – was added to the norms of desired femininity: the alternative radically crosswise gender formation and "importing" transverse meanings from the West. The exhibitionist way of talking about sexuality, displaying it, playing with it in art, advertising, press etc., transforming gender into an object of exhibition, violating gender in public scandalized the public. After fifteen years of such transformations nobody is shocked or disgusted at posters with nude bodies, acts converting gender into an instrument for different commercial purposes or abusing it. But is it thinkable to find a Polish D-Day, a striking event that brought changes to the gendered reality or (an) agent(s) in a time that made it possible to talk openly about human sexuality and avant-garde forms of conduct? I would like to argue that Poland did not experience any second wave feminist movement at all, as it was the case in Ireland. Under the communist regime, Poland had never had a public debate on contraceptives or even abortion, because the government allowed both, fully ignoring the Church's opinion on these matters. The sexual revolution of the 1960s and 1970s that took place in Western countries did not reach Poland, only its effects did get to the collective consciousness and brought about different forms of performance. This amounted to a kind of silent revolution, which the media only now expose, and disciplines such as sociology, women's and men's studies, gender studies and queer studies – analyze. Silencing the public voices before had resulted in transferring the discourse on issues around gender and sexuality into the private domain. In many cases the Catholic morality has not allowed an open discussion; paradoxically today's older generation frequently forgives the younger for their progressiveness, openness, readiness to accept democratization of all spheres of life, even in the most intimate domains. Agnieszka Graff, one of few Polish academic activists, argues that the Polish public is being told that patriarchal gender roles are firmly grounded in Mother Nature's plan (Graff: 2001). She attempts to demonstrate that while backlash rhetoric is an angry reaction to women's progress in the West, it is mostly just plain misogyny in Poland. In my view, this provocative statement does not fully correspond to the social reality. A much graver problem is the continuous direct or indirect arguing for desirable forms of femininity, women's need to be attractive, to be good professional, good lover, good housekeeper, good mother, good wife, and the simultaneous refusal to take women seriously when it comes to their gaining higher positions. The old ideal of a working mother in big cities became an ideal of a working single female without children who does gender by wearing extremely feminine clothes, by putting on make-up, by accepting flirts in the workplace etc. Another problem is that there is no place in Poland – a post-communist country – for anything that vaguely smacks of Marxism. The political culture in the post-socialist time is marked by deep aversion to even the slightest options inspired by Marxist ideals so that even social equality stays only a vague ideal that nobody dares to strive for.

In spite of the post-war strategies of doing gender "from the outside", high rates of employed women in Poland, a considerable number of well educated ambitious young women and men who, except for being professionally active, do get involved in the household, bringing up children etc. (Kula: 2004), some phenomena seem to be unchangeable and persistent in the public awareness. This can be focussed by means of language analysis, because each language emerges in a culture that itself values reality and forms it according to its settled norms. Language always esteems, assesses, and lexical choices take place often unconsciously. In the Polish context, for instance terms for professions that are grammatically feminine-marked have never been looked at too seriously. Polish women, who achieve professional success, do not fancy feminine forms for they have a trivializing effect. Why should these *significants* in their masculine grammatical forms make a better and graver impression of the *signifié*? The public perception of professions does in fact depend on the gendered suffix, to be – as a woman – a *lekarka* (female doctor) or a *prawniczka* (female lawyer) is definitely a less serious position than to be a *lekarz* (doctor) or a *prawnik* (lawyer). Only with few professions, particularly with these "prototypically feminine" ones, as teacher, nurse, shop assistant, secretary, charlady, tailor or singer, dancer, actress and designer, there is nothing bad about the feminine form (*nauczycielka, pielęgniarka, sekretarka, śpiewaczka, tancerka, projektant-ka* etc.), since femininity encoded in these jobs apparently vindicates their positive meanings and has to do with stereotypical codes of what is or can be called feminine and what masculine. When it comes to professions associated with public prestige, responsibility, proper earning and career perspectives, which have been conquered by women in the last two or three decades, one normally tends to use the masculine form to refer to both genders which projects a certain respect as e.g. *lekarz, prawnik, prezes*. The only turn in the linguistic gender-marking would be perhaps the example of *szefowa*, formerly a quite sarcastic term, and today a neutral female form for *boss* (derived from the male *szef*). Newspapers and magazines every day depict women in their professional positions but the media still take working women in well-paid high hierarchy positions as objects for a closer critical examination, as objects well suited for a whole topic of a feature. There is very little explicitness in the Polish media about men's problems, their lives in the public discourse which is a proof for the lack of emancipation in this field. In recent decades concepts of masculinities have experienced a transformation in Polish society, too and their prototypical codes have undergone enormous changes. Men are rarely shown as working fathers and husbands who express their intimate emotions in the media, even if the scope of representing men in the media is now much broader. The mediated space of privacy and working women does not go together with the formalized spaces which men occupy in the media. The public approval of plurality within gender roles and the rising acceptance of exchangeability of these roles make the Polish society slowly approach Western countries, as Kula (2004) emphasizes, but there is still a great need to work on practices of reproducing stereotypical notions and also on the distribution of positions in spaces of formal power in society, still extremely inequal.

The End of Irish Solutions to Gendered Irish Problems? The Birth of the "New Man"?

Whereas in Poland women were encouraged to work, plan their families and as a result of the totalitarian system society was turned into a matriarchy on the symbolic level, in Ireland women were, till the 1970s, legally prohibited to work in paid employment after getting married, to use contraceptives outside of marriage etc. There was no place for them to be individuals who have equal rights with men, equal pay and treatment in social issues, educational opportunities, adequate family housing. Politics was made and dominated by men[2]; family law, social and employment politics were done from the male perspective which always saw women as homemakers. Gender-oriented legislative process or "what the gender war is about" started extremely slowly only in the late 1960s. It was the Irish Housewives' Association that became a direct catalyst for setting up an ad hoc committee on women's rights in 1968. This committee achieved the establishment of the first national Commission on the Status of Women in 1970 which subsequently led to the formation of the Council for the Status of Women. Still there were many areas of social life and women's rights that were remarkably problematic from a gender equality perspective and solutions found to gender issues were only half-hearted. Married women were made into dependents, anti-agents in social welfare legislation which in concrete terms meant that until 1986 a woman was defined by her husband; women's pay was till the mid 1970s[3] much less than that of men. Women's movements of the 1970s campaigned for equal status in legislative practice in many areas. From the mid 1970s women's sexuality started to become an issue of public policy and social behaviour, the first Rape Crisis Centre was founded, thanks to the Health Act, sale of contraceptives to married couples, and later to persons over eighteen, was legalised. And only in 1992 – no more than because of the danger of AIDS – the purchase of condoms was extended to persons of seventeen years of age. A number of other legal standards has been entering the Irish legislation step by step, as women – having gained agency at last – were fighting for equal rights. This late liberalisation of norms still has effects today in either a quiet confusion about systems of values or radically polarized views on socio-cultural, moral as well as gender-related issues.

In the divorce referendum of 1986 the constitutional ban on divorce was removed, and the regulation was introduced that both spouses have joint

---

[2] Only in 1977 one could document some improvement in women's numbers being elected to the Dail. In 1993 there were already 20 women out of 166 MPs. The biggest success in the gender campaign was the election of Mary Robinson to the Office of President in 1990.

[3] Anti-Discrimination Act and Employment Equality Act.

ownership of the family home. Gender as a value throughout the 1980s was institutionalized, legislators became more sensitive to these topics, the bias against women in politics has been revised thanks to the women's movement that did not stop to campaign for women's rights, and thanks to the presence of women in the public arena. The state employment policies which for many years – exactly contrary to the Polish situation – inhibited women's participation in paid employment combined with the strong cultural and social influence of the Catholic Church, fostered a specific definition of the position of women in society and in the private sphere. In spite of some impressive statistics on the rise in employment numbers in the last years that Inglis (1998: 239) names and shows how new openings drew more women than men into paid work[4], there are still many more women than men in part-time employment (30 per cent of females to only 5 per cent of males), additionally working females are paid 15 per cent less than their male counterparts[5]. What is interesting in this context, is the fact that women are "locked into a handful of stereotypical areas" such as education, health, services (hotels or restaurants) still predominantly "enjoying" career options as waitresses, nurses or teachers. In more senior positions women still face obstacles and are marginalised; in all elite areas (civil service, business, government, politics, labour force and other leading positions) there is a constant under-representation of women, with females accounting for fewer than one-third of management positions. Irrespective of qualifications and competences, women face a "glass-ceiling" for promotion in many Irish organisations. Many spaces are still shut and in the search for remedies gender-related debates have been led by feminist critics.

In spite of these inequalities the Irish media show very well-educated women gaining more independence and professional chances. Women's views have become more liberal, their attitudes towards the Catholic Church – even if partly secularised – tend to stay more sympathetic than those of Irish men. Only a minority of Irish women still support the old conventional gender roles within the family while a noteworthy majority of men do stick to the "good old" but already outdated ideals and construct their masculinities around work, while those of women around the home as women's place[6]. In the public discourse

[4] "Between 1971 and 1996 the number of women at work rose from 212,000 to 488,000, an increase of 130 per cent. In the same twenty-five year period, the proportion of men at work increased by only 3 per cent" (Inglis 1998: 239).

[5] It means women are paid about € 4,600 less a year than men (*Irish Independent*, 08.03.2006: 21).

[6] A gripping study among men in the North-County Dublin was made by Goodwin (2002). The study shows that male identification of "real masculnities" is made through weaving around the notion of being male a chain of concepts such as "family, survival, support" (ibid: 162). Women in this study were described by men as "cheap secondary workers, mothers, housewives, passive, and dependent" (ibid: 163).

these issues get shaped differently: In the 1980s single mother was still stigmatized, today newspapers report on it as on one of possible lifestyles. The sexual revolution of the 1960s (in Ireland rather the 1970s – the 1980s) completed the sexualisation of women. *Spinster* disappeared as an evidently derogatory term and she could suddenly live her own independent unstigmatized life as a *single heterosexual woman* or *lesbian*. These broadened spaces of tolerance led to speaking freely of family-like unions, such as cohabitation, gay and lesbian unions, or one-parent-families. Such clear labels used in public discourse, before unthinkable, now became the order of the day. The drawback of the sexual revolution in Ireland, as it is the case in other European countries, is the pressure of the mainstream culture upon women in respect of their new roles as women: women are now quasi taught to be servicing male sexuality actively and not just passively. They must become agents in the sexual game and conquer spaces which before were reserved for men.

One side of the independence and freedom which employment possibilities have created for Irish women, is the capital that they dispose and that makes it possible for them to travel freely, invest a fortune in expensive nightwear etc. The image of a traditional modest Irish mummy with her robust figure who took care at home of her four or five children and her husband, who shied away from anything flashy or showy and could only cook well, has been transforming into the concept of a fashionable slim elegant sensual confident mum, who is a real agent in her life, of her life, who desires to be sexy, physically attractive and can afford trips to Malaga or Palma de Mallorca. After Irish women had gained agency in the private and the public, and got their hands on some money, they gave themselves an almighty makeover, doing gender quite offensively by conquering spaces that were forbidden to them before. The agency of wearing sexy alluring clothes, feeling free about sexuality, buying summer flats in the South of Europe started. Women began to symbolically colonize a crucial space that had been before reserved for men: decision-making and personal independence.

Masculinities in Ireland – like in other countries – have undergone an enormous change in recent decades. Heterosexual men had to make adjustments, accommodate to life with women in paid employment, take care of children, play caring fathers' roles, and reject oppressive forms of behaviour. They now have become "victims of women's liberation", as John Waters – a controversial *Irish Times* columnist – puts it. The growth of the gay liberation movement and the oppositional forms of masculinities it fosters have been remarkable developments, and already in the 1980s the media started to speak about the birth of the "New Men". New forms of masculinities: fathers, husbands, gay men appeared as a recurring subject in Irish media and Irish awareness a decade later, in the 1990s era of the "Celtic Tiger". As new labour divisions emerged, the institution of the family had to be revised and with that the male role of the wage earner. The functional space of breadwinning and providing the family with means to live, ceased to be a male domain only. Problems such as family planning and divorce, still in the 1990s solved in peculiar, context-oriented ways (*Irish solutions to Irish problems*), are not questions any more.

Agency encoded in masculinities, may have good and bad social effects: the image of men as those involved in crime, is one of the issues much discussed in the press nowadays. Since the beginning of the 1990s a big number of investigations (sociology, cultural studies, social psychology etc.) have sought to explain the link between maleness and crime (it suffices to look at crime statistics). These two criteria go hand in hand not because women do not take part in offending the law but because men in the public view became violent, less educated than women, unable to play their newly-defined roles as fathers, boyfriends, husbands of working women. This image of man as showing an anti-social behaviour and incapable to play the role of family breeder any more and symbolically castrated in his former roles, is in focus of much debate on masculinities that emphasize that they go through an identity crisis. The traditional male roles have been redesigned and rethought, their images, duties, attributes, responsibilities are no longer clear-cut and transparent. Some talk dramatically of "the dying phallus" (Clare, 2000: 3) since gender roles have rapidly transformed and the features of working agents and domestic "animals" got less and less clear-cut. Nonetheless, how O'Connell's (2001: 163-164) studies demonstrate, there are still surprisingly many men in Ireland who share the outdated views on gender issues, and cling to rigid and traditional gender roles. Maybe the male conservatism regarding these roles within the family can be interpreted as an attempt to metaphorically combat social changes and the inability to define themselves anew in the gendered order that has been for many years a result of the symbolic male dominance, theorized by Pierre Bourdieu in *La domination masculine* (2004 [1998]). Bourdieu suggests that within the framework of gendered power distribution and the obviously omnipresent male hegemony, certain knowledge produced constitutes the asymmetry of the male/female axis. In the process of decoding these structures, statements about women are traced back to their sex (which is supposed to determine them) and this approach seeks to demonstrate that characteristics referring to women do not transcend certain images corresponding with common social expectations as to what is typically feminine or typically masculine. Still now, women are portrayed in news media discourse mainly in their stereotypical status. This subordination within the male-dominated society is symbolically reinforced by sexist language, sexist advertising or certain strategies of representing women in their gendered specificity. Women are suddenly grammatically gender-marked, if the gender of the agent described in the news is female (*the female judge*).

Male-centered terms (*wife, girlfriend, mistress*) or the marital status, age, physical appearance are applied much more often referring to women than to men. Men are glossed by their professional designations or positions. These codified practices lead to depicting men as active, and females as passive, sexualized, even if women now are much more agents in their private and social lives than before.

Summing up the problems in today's Ireland, one may say that sex-category membership determines not only arenas of power, influence, recourses, privileges etc. but also new constellations of redefined roles in a society. The question is – if femininities and masculinities are public agencies, performances correlated with constant validation by the public – in how far we internalize these gender roles, strive to question them, and to what extent it is possible to challenge doing gender in these changing times.

Liquid Geography of Interweaving Gendered Old and New?

Today, in both peripheral countries by Central European standards, Poland and Ireland, women and men must choose between the old and the new rules of the game of doing gender, between the conservative and the progressive model, between extreme roles that are already inscribed in the kernel of these societies but which exclude each other and take their subjects in entirely different directions. In both cases the symbolic boundaries that keep categories "pure" and let societies be unique have already been – at least to some extent – broken. These transgressions of symbolic boundaries have introduced more obvious plural identities within the Polish and Irish societies, moreover, the pervasiveness of hegemonic masculinities as such has been called into question. With time, in both countries, the symbolic dimensions of alternative cultural codes have been immensely expanding and challenging traditional "old" lifestyles and their effects. Essentialist views on gender have been disappearing and it is becoming more and more difficult to uphold essentialist ways of perceiving females and males. The broadening public acceptance of variable forms of femininity and masculinity, the acceptance of certain hybrid forms and possibilities of altering one's doing gender during one's lifetime has become remarkable in both countries, even if the public tolerance emerges slowly and only moderately. Time has been one of the factors that did what it always does, even if agents bringing about these transformations (must) stay in the background, time changes models of life, and sooner or later brings – due to global changes – alternatives to accustomed concepts, codes of behaviour and ways of forming reality. However, in both countries, it is mainly women's rebellious agency changing these times that has brought about new definitions of gender, explored freedoms before hidden behind curtains of control, set new gendered spaces. Nevertheless, men and women in both locations still have a long way to go to reach spaces taken for granted in other geographies.

Works Cited

Bauman, Zygmunt (2003). *Razem. Osobno.* Kraków: Wydawnictwo Literackie.

Bauman, Zygmunt (2004). *Życie na przemiał.* Kraków: Wydawnictwo Literackie.

Bourdieu, Pierre (2004 [1998]). *Męska dominacja.* Warszawa: Oficyna Naukowa.

Butler, Judith (1999). *Gender Trouble. Feminism and the Subversion of Identity.* London & New York: Routledge.

Clare, Anthony (2000). *On Men: Masculinity in Crisis.* London: Chatto & Windus.

Foucault, Michel (1999). "The Incitement to Discourse". *The Discourse Reader.* Ed. Adam Jaworski & Nicolas Coupland. London & New York: Routledge, 514-522.

Goodwin, John (2002). "Irish Men and Work in North-County Dublin". *Journal of Gender Studies* 11.2, 151-166.

Graff, Agnieszka (2001). *Świat bez kobiet. Płeć w polskim życiu publicznym.* Warszawa: Wydawnictwo B.

Graff, Agnieszka (2003). "Lost Between the Waves? The Paradoxes of Feminist Chronology and Activism in Contemporary Poland". *Journal of International Women's Studies* 4.2 (Special Issue: "Harvesting Our Strengths: Third Wave Feminism and Women's Studies"). <http://www.bridgew.edu/SoAS/jiws/April03/Graff.pdf>.

Inglis, Tom (1998). *Moral Monopoly. The Rise and Fall of the Catholic Church in Modern Ireland.* Dublin: University College Dublin Press.

Kula, Magdalena (2004). "Matka Polka wyzwolona". *Gazeta Wyborcza*, 27.04.2005.

Marody, Mira & Anna Giza-Poleszczuk (2000). "Changing Images of Identity in Poland: Form the Self-Sacrificing to the Self-Investing Woman?" *Reproducing Gender. Politics, Publics, and Everyday Life After Socialism.* Ed. Susan Gal & Gail Kligman. Princeton, New Jersey: Princeton UP, 151-175.

O'Connell (2001). *Changed Utterly. Ireland and the New Irish Psyche.* Dublin: The Liffey Press.

Rose, Lotte (1997). "Körperästhetik im Wandel. Versportung und Entmütterlichung des Körpers in den Weiblichkeitsidealen der Risikogesellschaft". *Ein alltägliches Spiel. Geschlechterkonstruktion in der sozialen Praxis.* Ed. Irene Dölling, & Beate Krais. Frankfurt am Main: Suhrkamp, 125-152.

Tovey, Hilary & Perry Share (2000). *A Sociology of Ireland.* Dublin: Gill and Macmillan.

West, Candace & Don. H. Zimmerman (2004). "Doing Gender". *The Gendered Society Reader.* Ed. Michael Kimmel. New York & Oxford: Oxford UP, 150-168.

Orly Benjamin & Sarit Sambol

Surviving Spatial Oppressions over Time

Israeli Working Poor Women's Localized Resistance

Introduction

Travelling forms of feminism may produce hybrid cultural formations of indigenous feminism in their encounter with local social structures and biographies. Feminist projects that are directed at women working together to resist power structures, may be localized by an enhanced spirit of women supporting each other but they may also loose some of their spirit in their encounter with local multi-generational forms of power and poverty. Following Aptheker's view of women's resistance as dependent on their resources, we raise an empirical question regarding an economic empowerment feminist project which was imported to Israel by feminist activists: to what extent are travelling forms of feminism able to maintain their commitment to women working together, vis-à-vis spatial and temporal oppressions? We analyze 12 stories of localized Israeli working poor women's resistance in order to trace the struggle between continuing forms of isolation and attempts at connecting and working together with other women.

Feminists, since long, resisted the hegemonic disparity between the personal and the political. Focusing on women's resistance, particularly that which emerges within the framework of feminism guided empowerment workshops, requires a discussion of spatial and temporal dimensions of agency and resistance. The spatial dimension refers to the ongoing negotiation which consolidates our identity within the specific localities in which we operate. The temporal dimension refers to the potential process through which we are able to accumulate knowledge, develop awareness and undergo transformations in terms of our daily behavior, thought and talk (Benjamin, 2003). Both the spatial and temporal dimensions are of course central to identity development, identity politics, and the way identity connects to social movements. It is our identity that connects us to the feminist movement and simultaneously our identity is fed and constituted within the feminist movement's discourses and practices primarily by the reinforced possibility to reject hegemonic frameworks of meaning and resist them. When this process takes place to the extent that institutional, interactional or individual mechanisms reproducing hegemony are challenged and resisted, we name it empowerment (Ward and Mullender, 1991).

Economic empowerment projects are a particularly interesting case in that they consist of feminist projects that are directed at women working together to resist power structures. The boundaries between personal and collective resistance are blurred because such projects encourage an enhanced spirit of women supporting each other. These projects generate a unique encounter between feminists and other women. Feminists are women who have already developed feminist awareness and share the understanding that the given social order privileges the masculine and distributes power inequitably according to gender (Stanford Friedman, 2000). Moreover, they are determined to put their efforts and resources to the improvement of women's quality of life which is damaged in the context of the gender, race and class power structures. Participants in such projects do not necessarily share these perceptions or determination. Hence, the empowerment process may potentially serve as an opportunity to invite them to recognize the spirit of women working together, as a resource. In this context, empowerment transforms and becomes a form of intervention on both the spatial and temporal dimensions of agency and resistance. This specific process that takes place in the encounter which is generated by the economic empowerment workshop, is examined in this article as an opportunity to theorize on the personal and political working of women's resistance.

In the opportunity formed within the economic empowerment workshop, an encounter is generated between a travelling form of feminism that was born in the USA and in the third world, one that is focused on sustainability which was foreign to Israeli reality during years of a locally developed welfare state, and a specific local stratified reality. Namely, local multi-generational forms of power and poverty, that had taken the shape of specific spatial oppressions in the Israeli periphery and slum neighborhoods, and geographical areas in which poverty had dominated for lengthy periods of time and in which various resistance activities have been organized over the years. Women in these geographical areas often carry poverty as a critical temporal characteristic in their habitus, being the daughters of poor families, struggling against poverty in various life stages. These spatial and temporal dimensions may confine resistance to focus specifically on an interactional level rather than wholly enabling the institutional challenge which significant economic empowerment would entail. Here we raise the possibility of unifying the personal and the political through analyzing the emergence of resisting ways of talking and thinking that become central in connecting more political and collective forms of resistance to more personal and individual acts. We aim here at examining the possibility that travelling forms of feminism, those arriving from global trends of dealing with women's poverty in developing countries, may produce hybrid cultural formations of indigenous feminism in their encounter with local social structures and biographies. In our attempt to portray the personal and the political of resistance as a continuing sequence rather than fragmented isolated acts, we adopt the insight that "if hegemony refers to that which is unthinkable, resistance must depend at some point on thinking the unthinkable" (Ewick and Silbey, 2003). We set out to examine the possibility that within the encounter between the global and the local of women's economic empowerment workshops, a place of thinking the unthinkable is generated. In other words, we ask whether the personal and the political of resistance are bridged trough raising the possibility that organized feminists are facilitating the possibility of less organized women, to think the unthinkable.

The Personal and the Political of Resistance

Several authors have already attempted at presenting discussions that undermine the dichotomy between the personal and the political by conceptualizing everyday survival experiences, as resistance. Aptheker (1989) in her discussion of poor women's resistance as expressed in everyday experiences and struggles for survival, makes this point of personal resistance bearing political meanings, despite the lack of collective action characteristics. She further explains that resistance is "about creating the conditions necessary for life, and it is about women expanding the limits of the restrictions imposed upon them" (169: 1989). In other words, Aptheker emphasizes the spatial dimension of resistance in her pointing out to boundaries and restrictions as the form of women's oppression. Because of structural boundaries women are excluded from public sphere's "valuable" activities and thus remain isolated particularly in the context of dominant individualistic ideologies. Isolation is the outcome of structural exclusion and limited access to resources. In its turn, isolation limits women's ability to reinforce their access to resources. Thinking of the oppressive ways in which isolation operates in women's lives, actions that challenge isolation can be seen as expressions of women's resistance. In this context, the challenge of isolation becomes possible when women gain some access to resources and only when women manage to challenge their isolation, they come to the position from which they can facilitate their accumulation of resources. Thus, Aptheker's (1989) view is one that is concerned with the spatial and temporal dimensions of resistance: the ability to expand social boundaries, an ability that is transformed over time and shifts between ways of perceiving oneself as well as understanding the world, is central to women's resistance. And the importance of the expansion of one's boundaries is not reduced by the absence of challenge to macro level structures of power.

In her account of the literature on women in poverty and their acts of resistance, Lister (2004) explains that mostly resistance in this field is equated with unrecorded work, dealing with welfare authorities, and the rejection of negative labelling. These expressions of "everyday resistance" correspond with Scott's suggestion (1985) that even the powerless have weapons to use against the privileged even if the goal of using such weapon is not political change but rather, mere survival.

Moreover, resistance in poor women's lives attracted attention as a way to depart from traditional and inferiorizing ways of treating women in professional texts. Thus scholars attempted to study "[...] the ways in which women handle social arrangements that discriminate them- how women act to oppose them, how they design their actions in ways which suit them and their daily social context and finally, which of the anthologizing social concepts attributed to them do they absorb" (Kromer-Nevo, 2000: 50).

Survival efforts, creating a protective space for children as well as widening the perceived range of open life-style choices as forms of everyday resistances have in common two general denominators: firstly, they do not necessarily require an independent definition of the situation, one that would challenge hegemonic definitions and secondly, they do not require collective action. They are all expressions of humane agency that is perceived as making private decisions and acting on them with minimal co-operation of others. Importantly, they do require knowledge, familiarity with a wide range of possibilities and the courage to suffer the prices of social control whether sporadic or state activated. An anthropologists view of the current social control exerted upon the poor name it "stigmatization" (Munger, 2000). Social control often deploys shame in ways that exclude and stigmatize the poor even when not actively engaging in resisting action. Shame, as Scheff (2003) explains, exerts social control through exclusion and isolation. Shame is a common emotion in an ideological context that is fast to blame the poor woman with her poverty. In this context, shame is a predominant measure of isolation. The options opened for the poor to escape shame are confined and further confine the social space in which the poor are protected of shame.

The Israeli Organization for Economic Empowerment

Up until 1985, the year in which globalization was formally adopted through a vast economic reform, decommodification was central to Israeli employment arrangements and welfare policies alike. Thus, poverty takes a local form to which growing gaps are central but some access to social rights still exists. Among the citizens, Israeli-Palestinians benefit of the most limited access to social rights. Immigrants from Arab countries (Mizrahi) and their children and sometimes grandchildren, follow them in their limited access to social rights. And finally, immigrants from the Former Soviet Union are often trapped in poverty, while making efforts to enter the local labor market with its current neo-liberal non-standard shape. Further, those living in the periphery of the country in what is called development towns, as those in slum neighborhoods, face limited opportunity structures.

In Israel, social policy, according to Fogiel-Bijaoui (2003), is now mostly equated with poverty, while failure to "make it" in the labor market is mostly explained in terms of "ethnic backwardness" and/or as the result of "personal distress". She further explains that the change from universal state-provisions to means-tested provisions marginalizes social rights and leaves women, who are second wage-earners and/or first care-givers, as second class citizens. Social rights allocation has thus become arbitrary and has been turned into charity. The neo-liberal discourse currently hegemonic in Israel turned women's access to social rights minimal. Similar to the way globalization has had its impact elsewhere, next to the sharp restructuring of welfare policy, high unemployment rates are maintained (in Israel, the later are always higher for women). Taken in the context of gender, unemployment rates operate also to reinforce traditional perceptions of women's roles (where women's employment is sometimes presented as the reason for men's unemployment). Women's dependency on welfare and their vulnerability in their intimate relationships are also supported by unemployment both through the blockage of independent income and through the proliferation of non-standard employment. Thus, unemployment rates operate in mal directions not only in relation to unemployed women but also to those inside the labour market by generating forms of cheap employment: jobs that are harmful and precarious.

In Israel, the economic empowerment approach was adopted by the "Economic Empowerment for Women Association", which set for itself a goal of helping women living in economic and social stress to step out of poverty. The course, which is conducted by the Association, includes four stages: recruiting the participants; an economic empowerment training workshop in which ten meetings aim at personal empowerment and ten meetings aim at facilitating a business initiative; finally, the fourth stage is handing out a small loan ranging between 500 to 2500 dollars for the purpose of applying the business plan. After a woman has taken the loan she is guided for a year by meetings and personal counselling. The main criterion for participants' recruitment is the candidates' inability to take a loan from a regular commercial bank because they do not own property or receive a salary that can vouch for them financially.

The organizers of the project in Israel see it as a feminist project. They have been active members in a feminist centre in Haifa for about twenty years. They see the project as emerging from their feminist prolonged activity and the outcome of their realization that poverty is going to become a lot more prevalent and that it is crucial to women's ability to be protected from violent relationships as well as to their ability to use health care services and educate their children. As feminists, they hold a realistic perception of the project and talk explicitly about the long and difficult process that is involved in actually escaping poverty. However, they see their goals as smaller as well. For example, one important aspect is facilitating the transformation between the commonplace gift economy and the money-based economy, thus encouraging women to require payment for work which they do as favours or as gestures of friendliness. Another important goal is that the group who has gone through the process together will maintain its collective basis and will turn into a forum of reciprocal support and consultation. Moreover, the leaders of the projects take pride in the fact that other organizations do a similar kind of work, but they primarily focus on those women whose weak social position prevents them from using these types of services when given elsewhere.

Potentially speaking, a project of economic empowerment could expose women to ways of talking and thinking which they have not encountered before. Further, the workshop could become an opportunity to think the unthinkable and to share it with others. Put differently, we focus on localized resistance understood as the process in which an individual accumulates resources to the extent of firstly, voicing what is silenced by structural hierarchies generating her economic position and secondly, speaking and acting in ways that attempt at challenging her social isolation, shame in her world and her economic position.

Methodology

Following feminist researchers, we concentrated in this study on women's experiences attempting to understand them in broader social contexts (Reinhartz, 2001). This understanding is accompanied by the notion that the researchers as well as the subjects are a part of the research process, and the "story" being told in the research itself is the outcome of the interaction between them (Edwards, 1993).

Twelve women, who reside in an inner city neighborhood located in one of the biggest cities in the country, were interviewed for this study. We reached the women as part of a longitudinal study that dealt with economic empowerment among Israeli poor women. All the interviewees participated in an economic empowerment course organized by the "Economic Empowerment for Women" Association.

The women were interviewed both at the early stages of the workshop and after it ended. The semi-structured interviews (Berg, 1992) focused on women's life stories as well as their experiences at and expectations from the course. Content theme analysis (Berg, 1992) enabled us to undertake both a deductive and inductive perspective using emerging themes such as "coming out" as well as themes we were interested in such as change and resistance.

Findings

Isolation is one outcome of power structures which elicits in individuals self blaming and shame to the extent that they are silenced (Scheff, 2003). Within gender, race and class based power structures exclusion and isolation are simultaneously reinforced by external forces and inner feelings. People of oppressed categories experience emotional practices of othering (Schwalbe et al, 2000) and inevitably learn to operate within separate spheres.

The process of othering is somewhat challenged in women's experiences during the economic empowerment workshop. The transformation can be generally described as a shift between isolation and working together with others. The transformation is rarely directly attributed to the economic empowerment workshop. Instead various resistance activities are counted as central to the transformation. The need to resist isolation is the one most often articulated in the interviewees' explanation of why it is that they couldn't use the workshop for the economic level of coming out.

> S: What do you mean when you say mobilize things?
>
> Y: If you take me, I needed a mentor and I didn't have such a meeting with a mentor. Then, they say you could no longer have a meeting with a mentor so I didn't have one. I need someone to take me and tell me: listen, that is the women who join you and you begin, I don't yet know how to build good models and that [she refers to the specific field in which she would hope to open her small business - sewing], to begin build a model, and you will start mobilize it. (Yael Avraham, 60, self-employed).

Yael Avraham explains that to begin, to mobilize, to get out, she needs a partner. Somebody who would undermine her isolation and will share the risk with her. Elsewhere in the interview she explains that it's a problem with her "personality", that she's unable to take the risk. She'd rather join someone else in her business than face the risk of having to return a loan after getting stuck with the wrong kind of clothes. Her interview reflects how lonely she is vis-à-vis the authorities, always having problems negotiating her rights with them. Her understanding of the workshop is that her isolation was not broken. She is still on her own, frightened and stuck. Her case is one in which loneliness has remained solid. No encouragement was found for her in the workshop to connect to other women in ways that will empower her. Despite her feelings that as a former community worker she has been always interacting with other women, always spreading around knowledge, encouraging women to defeat their faith in harsh physical work, she herself remains isolated. When the workshop ended she was left feeling that her strength and inner resources could not allow her to locate a source of significant support. Here, on the basis of Yael Avraham's experience, the local nature of the economic empowerment workshop seems to defeat its purposes. As explained by the organizer of the Israeli organization for economic empowerment, adopting the micro-finance program from third world countries into developed countries required some changes. The main change was the turn to individual loans from collective loans. A spatial dimension is revealed: Israel is perceived as a first world country hence the assumption of individualist responsibility and economic activity. This assumption meant that women are not encouraged to join together their projects and risks involved. The isolation is adopted into the project itself by expecting women to connect and develop courage on their own.

Economic empowerment had happened for some of the participants despite the local nature of the rebuilt isolation. This happened to those who resisted the tendency to forget and who were determined to adopt the messages of the workshop, at least partly through maintaining relationships with individuals who attended the workshop. Like in the process of learning a language, holding on to people who speak the language learned at the workshop may turn resistance into a more significant experience. Apparently the messages significantly entered the mind of Marinna Levi and consequently she experienced herself as climbing a step up.

> First of all, it was a real change of life. A stage in which I climbed another step up... And it was for my own good both the workshop and the organization. ... I grew up. It's as if I grew up.
>
> S: Would you give me an example? What do you mean when you say that you grew up?
>
> M: First of all, the knowledge. I didn't know how things work and I didn't know what to begin with. After I finished the workshop, first of all, I knew. Now, all the material is accumulated and I can go through it and refresh… all I need to do is take out the papers and simply look and see what seemed at the time important. So, any knowledge is a step upward and another grade up.

The temporal dimension of the empowering effect of the accumulated knowledge is revealed in the way Marinna Levi understands the shifts in her experiences. She too, like other participants, feels that she forgot much of the knowledge that she was exposed to at the workshop. However, unlike others, she still sees the knowledge as a resource: "and I can go through and refresh". She takes responsibility and celebrates the possibility to refer to the notes she took during the workshop which seem like an empowering resource. She continues:

> In fact... it helped me in my work too... I started to talk to people who had a small business... at that time, they wanted to fire me but because I knew better how to run a business and I gave our boss the small advice I got at the workshop... it looked really good. They felt I care about their business as if it was my own and they kept me there and I feel I became better at my work particularly in negotiating with clients because of these small tips.
>
> S: Had the workshop changed the way you perceive yourself or how you think of yourself?
>
> M: Yes... I recently began to open my mouth more and speak and say things then suddenly I feel I'm important to people and they want to ask me and this is pleasant and gives me the confidence. The main thing is the confidence. (Marinna Levi, 35, Kindergarten owner).

Resistance takes the shape of becoming more of a normative person, managing to hold a job for longer, accumulating knowledge, developing relationships and a more positive self-perception. The empowerment workshop paves the way towards resisting one's economic faith by opening up options that were blocked before. And indeed, after the interview, Marinna Levi contacted Sarit to inform her that she had opened up a kindergarten of her own for under 4-year-olds and that she manages to earn significantly better than she used to earn at the shop where she worked before. What seemed to be a simple step outside her isolation, her relationship with a friend who planned to open such a kindergarten, became in the context of the workshop a real plan of action that fills her up with positive feelings about herself. The cycle of an increased sense of self-efficacy began to work for her allowing her to receive better feed-back and continue to grow.

Discussion

Working poor women in vulnerable social categories, both, primary and secondary providers, bear in their biographies the scars of their spatial oppressions, those shaped in the context of immigration and the national conflict. These women arrive at the economic empowerment workshop with the hope of increasing their economic activity and the rewards it embeds. All the interviewed participants, both those who managed to increase their economic activity and the others, remain poor working women after the workshop. The short term character of the workshop and perhaps also its somewhat individualistic character, left their poverty, the continuous existential worries, untouched. The knowledge, the language, the skills they learned could overcome neither their spatial oppressions nor inter-generational transmission of poverty. The workshop leaves women to face on their own the temporal dimension that is expressed through the possibility to compare between past, present and assessed-future opportunities structures. No new understanding of their oppression over time is revealed in their interviews. However, an important empowering process is taking place.

The feminist movement in Israel, the local feminism, has been mostly indifferent to issues of poverty and working conditions focusing rather on equality, violence against women and sexual harassment. Consequently, the feminist movement in Israel has found itself trapped in small circles avoiding direct encounter with women in poverty. Several organizations benefited of a relatively homogeneous composition which meant that Israeli feminists have found it difficult to cross boundaries of ethnic, religious, political, class and national identities.

The economic empowerment workshop is unique in that respect. It really managed to cross a significant social distance and get in touch with underprivileged women who are rarely seen in feminist activities and gatherings. For these women the encounter with the feminist activists through participating in the economic empowerment workshop touches upon issues of shame. The workshop created a sphere in which women could tell their "being powerful" stories, a telling which turned the shame of poverty into the pride of survival. At the same time, we learned from the second wave of interviews, which were conducted a few months after the workshop was over, that a new type of shame was there for the women to struggle against. This new type of shame was connected to the sense of failure which quite a few women among the participants were now feeling vis-à-vis their lack of ability to undertake the challenge of the micro-credit business.

The first interpretation that we raise in our discussion is that we failed to trace the foot prints of a feminist discourse in our interviewees' ways of talking. In no case we encountered in our analysis a feminist slogan or statement. The footprints that we did encounter are those of the self-development individual growth discourse together with individualistic understandings of one's own responsibility for poverty. This finding is consistent with existing feminist criticism of micro-finance programs.

The resistance we traced left social structures and their relevant ways of shaping biographies outside the ongoing discussion along the workshop. The economic empowerment workshop became a feminist project of the hybrid type of being directed at women working together to resist their economic faith. Clearly, it had been localized by an enhanced spirit of women supporting each other. However, in its encounter with local multi-generational forms of power and poverty, those which survived over time and political changes, this feminist project lost some of its global spirit. Particularly the move between individuals working alone to collectives working together, was dismissed. Rather than encouraging women to enter collaborations with other women, the program is designed to encourage individual projects which remained too frightening for some of the participants. In this sense, local mechanisms of isolation were left unchallenged.

Following Bettina Aptheker's (1989) view we found in our interviews stories of resistance that were scattered along women's life stories. Resisting local authorities aiming at separating them from their children repeatedly emerged. Smuggling under conditions of scarcity was described by one of the interviewees and a few described forms of community activism. This clarified the extent to which resistance is always relative to the tools and resources women have available (1989: 180). Moreover, it clarified that the economic empowerment workshop had modified women's perceptions of their resources only in very few cases. Fear and shame were left too vivid in the others' experiences. Nevertheless, the economic empowerment workshop's strength was in facilitating the accumulation of interactional and individual resources. It sometimes supported participants' ability to tackle the internalized fears of being and acting themselves. In enabling women to meet each other, to tell each other their stories of resistance as well as sharing their fears and inner doubts, a few women were able to form long standing relationships that helped them resist the power of isolation. In this sense too, Aptheker's view sheds light on women's struggles vis-à-vis the power structures that shape their lives.

We found that women's resistance has been "about creating the conditions necessary for life, and it is about women expanding the limits of the restrictions imposed upon them" (169: 1989). If isolation is a restriction that is imposed on poor women, then resisting isolation has a political significance which challenges the disparity between the personal and the political of resistance. Our question regarding resisting spatial and temporal oppressions met empirical evidence which pointed out the temporal dimension of women working together as central to the understanding of the workshop's ability to facilitate resistance to spatial oppressions. Our interviewees clarified in their accounts the enormous surge of energies and positive empowering ways of thinking which they had experienced in the primary stages of meeting with the other women. A powerful spirit of women working together was remembered by the participants. However, the temporal dimension was more powerful in that once the workshop was over, women hardly remembered, could hardly connect to this spirit and were unable to draw power for themselves from it. Their daily struggles to protect their kids from the shame of poverty would necessarily continue to be scattered with instances of local resistance.

Works Cited

Aptheker, Bettina (1989). *Tapestries of Life: Women's Work, Women's Consciousness and the Meaning of Daily Experience.* Amherst: University of Massachusetts Press.

Benjamin, Orly (2003). "The Power of Unsilencing: Between Silence and Negotiation in Heterosexual Relationships". *Journal for the Theory of Social Behavior* 33.1, 1-19.

Berg, Bruce L. (1998). *Qualitative Research Methods for the Social Sciences.* London: Viacom.

Edwards, Rosalind (1993). "An Education in Interviewing: Placing the Researcher and the Research". *Researching Sensitive Topics.* Ed. Claire M. Renzetti, & Lee M. Raymond. London: Sage, 35-60.

Ewick, Patricia & Susan S. Silbey (2003). "Narrating Social Structure: Stories of Resistance to Legal Authority". *American Journal of Sociology* 108.6, 1328-1372.

Fogiel-Bijaoui, Sylvie (2003). "Familism, Postmodernity and the State: The Case of Israel". *The Journal of Israeli History: Politics, Society and Culture*, 21.1-2, 38-62.

Kromer-Nevo, M. (2000). *Life Stories of Women Who Live in Prolonged Economical and Social Distress in Israel.* PhD in Philosophy. Jerusalem: Hebrew University. (Hebrew).

Lister, Ruth (2004). *Poverty.* Cambridge: Polity Press.

Munger, Frank (2000). *Laboring Below the Line: The New Ethnography of Poverty, Low-Wage Work, and Survival in the Global Economy.* New York: Russell Sage Foundation.

Reinharz, Shulamit (1992). *Feminist Methods in Social Research.* New York: Oxford UP.

Scheff, Thomas (2003)."Shame in Self and Society". *Symbolic Interaction* 26.2, 239-262.

Scott, James C. (1990). *Domination and the Arts of Resistance: Hidden Transcripts.* New Haven: Yale UP.

Schwalbe, Michael L., Sandra Godwin, Daphne Holden, Douglas Schrock, Shealy Thompson & Michelle Wolkomir (2000). "Generic Processes in the Reproduction of Inequality: An Interactionist Analysis". *Social Forces* 79.2, 419-453.

Stanford Friedman, Susan (2001). "Locational Feminism: Gender, Cultural Geographies, and Geopolitical Literacy". *Feminist Locations. Global and Local, Theory and Practice.* Ed. Marianne DeKoven. New Brunswick: Rutgers UP, 13-36.

Ward, Dave & Audrey Mullender (1991). "Empowerment and Oppression: An Indissoluble Pairing for Contemporary Social Work". *Critical Social Policy*, 11.2, 21-30.

Agnieszka Kołodziej-Durnaś

Time, Space and Agency in a Life Project

Polish Students' Reflexivity

Introduction

In recent years Poland has undergone major changes in its political and social system. Initially, this was possible thanks to some actors' awareness of agency potential and in return it must have further affected the Polish mentality and individual ways of thinking about the future. The case of Poland is special as the changes in creating a self-identity and designing one's own biography are influenced not only by high modernity (com. Giddens, Pierson 1998: 118) or postmodern times (as others prefer) but also by system transformation mechanisms. The topic is very broad and only some of the mentioned phenomena and the processes' aspects can be addressed here.

Analyses of contemporary Polish society often comprise a statement on its enlarged reflexivity and quality of constant self-reference which result in intensive individual activity (com. Koralewicz, Ziółkowski 2003: 244-245). It seems that Polish students constitute an exceptionally interesting category of respondents to find out how people incorporate the issues of time, space and agency into thinking on their lives. First of all, this results from their being Polish – this fact combines the question of the situation in post-communist countries with the notion of globalization processes. Secondly they are young – their way of constructing their own lives will probably influence the future of the rest of the population more than the way others construct them. Thirdly because they are members of a society which has just joined the EU preceded by a long reflexive period of negotiating terms of accession and adjusting the state's institutions and regulations to European requirements. And finally – last but not least – members in this category may be perceived as carriers of some qualities, skills and attitudes formerly associated with the stratum of the intelligentsia (intellectuals). This stratum is now becoming more and more similar to the class of professionals (common for Western states) who have a different identity and perform other functions in society.

In this paper the ideas of a reflexive monitoring of action, agency and life politics are discussed. A review of contemporary authors' thoughts on the issue is followed by the results of a preliminary research. About 200 students from the University of Szczecin and other higher education institutions in this city were surveyed in the spring of 2004 just before Poland's accession to the EU. The respondents were asked among other things whether and how they plan their future, whether they plan to leave Poland, whether they would accept others leaving Poland, whether they feel they have an influence on their lives, whether they have long or short term goals and whether they perceive themselves as patriots. They were also asked to predict the 'factors' or 'parts' which would play a role in their future life for the next 10 years. All the responses were intended to find an answer to the question about how students incorporate time and space in their agency and reflexivity concerning their biography and whether they treat it as their own life project.

Wide Cultural Context

After 15 years of transformation Poland more and more resembles the major states of Western Europe. Western civilization has been an example of individualist culture for a long time as it is arguably definitely less collectivist than some Asian countries. Geert Hofstede carried out a huge research project and worked out the detailed characteristics of cultures and their analytical dimensions. He outlined key differences between collectivist and individualist societies in terms of general norms, family, school, workplace, politics and ideas. Here are only some of the differences: children learn to think in terms of 'I', not 'me', identity is based in the individual not in the social network to which one belongs, the purpose of education is learning how to learn and not how to do, everyone has a right to privacy and is expected to have a private opinion. The ultimate goal is not harmony and consensus in society but the self-actualization of every individual (com. Hofstede 1991: 67, 73).

It seems that in times of globalization individualism as a cultural feature is on the increase. In the social sciences it is manifested through an incredibly increased number of publications on 'identity' (com. Bauman 2001: 140). Also the notions associated with the individual multiply. One of the most significant here is 'life politics': 'the politics of self actualization in the context of the local and global and the emergence of the internally referential systems of modernity' (Giddens 1991: 243).

As the world (especially the Western world) is changing so rapidly the freedom of the individual is both more real and more scary. As Zygmunt Bauman points out there is always a link between freedom and anxiety:

> [...] the discovery that things do not keep their shape once and for all and may be different from what they are is an ambiguous experience. Unpredictability breeds anxiety and fear: the world is full of accidents and surprises, one must never let vigilance lapse and should never lay down arms. But the unsteadiness, softness and pliability of things may also trigger ambition and resolve: one can make things better than they are, and need not settle for what there is since no verdict of nature is final, no resistance of reality is unbreakable. One can now dream of a different life – more decent, bearable or enjoyable. And if in addition one has confidence in one's power of thought and in the strength of one's muscles, one can also act on those dreams and perhaps even force them to come true... Alain Peyrefitte has suggested that the remarkable, unprecedented and unique dynamism of our modern capitalist society, all the spectacular advances made by 'Western civilization' over the last two three centuries would be unthinkable without such confidence: the triple trust – in oneself, in others, and in the jointly built, durable institutions in which one can confidently inscribe one's long-term plans and actions (Bauman 2001: 141-2).

Confidence and trust on one hand and on the other the permanent need for constant monitoring of the self and the social environment are the reasons why reflexivity is so important in the individual as well as the collective life, maybe reflexivity is the *sine qua non* condition of agency.

Reflexivity, Agency and Life Projects

Reflexivity and agency are the aspects of (post)modernity most often discussed by many contemporary sociologists. Let me just mention Anthony Giddens, Ulrich Beck, Jürgen Habermas – although they sometimes use slightly different notions, they all speak about similar phenomena. One probable thesis stemming from their texts could be as follows: modern human beings construct their biographies reflexively and so do companies and states/nations. Or to use Giddens', Archers', Elias', Bourdieu's perspective – there are no separate levels of social existence: actors, organizations, macro-scale. The social process of structuration is permanent and reflexive, actors both use and create rules and regulations constituting structure or rather the process of structuration. The following terms, defined by Giddens, are associated with the topic:
Reflexive project of the self: the process whereby self-identity is constituted by the reflexive ordering of self-narratives (Giddens 1991:244).
Self-identity: the self as reflexively understood by the individual in terms of his or her biography (Giddens 1991:244).
Trajectory of the self: the formation of a specific lifespan in conditions of modernity, by means of which self-development, as reflexively organized, tends to become internally referential. (Giddens 1991: 244).
Colonisation of the future: the creation of territories of future possibilities, reclaimed by counterfactual interference (Giddens 1991: 242).

Narrative of the self: the story or stories by means of which self-identity is reflexively understood, both by the individual concerned and by others (Giddens 1991: 243).
Life-planning: the politics of self-actualisation in the context of the dialectic of the local and global and the emergence of the internally referential systems of modernity (Giddens 1991: 243).
Reflexive monitoring of action: The purposive, or intentional, character of human behaviour, considered within the flow of activity of the agent; action is not a string of discrete acts, involving an aggregate of intentions but a continuous process (Giddens 1984: 376).

Giddens and the above mentioned authors among others take part in the debates called agency versus structure and challenge the dualism of macro versus micro and individual volition and collective oppression/superiority. That is how the issue of agency is perceived in contemporary sociology:

> *Agency* – the term is usually juxtaposed to structure and is often no more than a synonym for action, emphasizing implicitly the undetermined nature of human action, as opposed to the alleged determinism of structural theories. If it has a wider meaning, it is to draw attention to the psychological and social psychological make-up of the actor, and to imply the capacity for willed (voluntary) action (Marshall 1998: 10).

Agency and reflexivity, apart from being crucial notions for contemporary social sciences, also serve as crucial phenomena for the future shape of the global existence. The issue is not only the topic of the latest books by great social thinkers but is also present in major scientific magazines in the field of social sciences. John W. Meyer and Roland L. Jepperson are the authors of one of these papers. They emphasize the significance of culture in the process of constructing agency as well as the fact that the modern actor becomes an authorized agent thanks to culture and not as disembedded from culture (com. Meyer, Jepperson 2000).

The reflexivity and agency of students as a more specific theme has also been discussed by many authors worldwide. Many aspects have been featured and elaborated on the situation of contemporary youth. Devine Nesta and Irwin Ruth analyse the individual and its autonomy. They use notions like 'technologies of the self' and try to examine how body and mind, individual and environment, thoughts and behaviour diverge in the process called consciousness. They also outline some implications for teachers and students within the school context (com. Devine, Irwin 2005). Julie McLeod constructed her article in relation to Anthony Giddens' concepts amongst others. She elaborates on reflexivity, detraditionalisation, various ways of constituting and knowing the self. The author also suggests that the transformation of intimacy somehow enforced by this excessive reflexivity triggers the establishment of gendered subjectivities. But the possible conclusion which might be drawn from this article seems to go beyond gender limitations: the reworked intimacy also refers to (and is a part of) an individual biography which is better understood as life politics (com. McLeod 2002). In one of her other papers the same author attempts to present prospective and retrospective understandings of identity over time and forms of subjectivity (com. McLeod 2003).

Colin S. Cremin focuses on another facet of the phenomenon of reflexivity. The author examines documents schoolchildren routinely complete to be presented to potential future employers. He illustrates how the data that the pupils insert are adapted, reflected upon to make oneself suitable to the current labour market. It seems that from the very beginning of their youth students are forced to be reflective but also to construct their biography project not only according to their needs but also to the requirements (or perceived as such by young social actors) of the environment (com. Cremin 2005). On the other hand Garrett J. McAuliffe pointed out the fact that the experience of professional preparation additionally increases reflexivity among students, as well as their autonomy, opportunities for independent thinking, and capacity for dialogue (com. McAuliffe 2002). Similar results (higher academic self-confidence, improved social agency and a disposition for critical thinking, development of certain aspects of the self) are caused by more experiences with diversity, especially interactions with diverse peers. This might mean that students who are planning to leave are at the beginning both more reflexive as well as critically and creatively oriented in their way of thinking but will also be in an even better situation in terms of possibilities to develop their reflexivity in the future (com. Nelson Laird 2005)

Polish Students and the Academic World

In Poland as well as in other countries of the post-communist block students traditionally constituted the future elites called the intelligentsia. The main role of an intelligentsia member was not only to deal with knowledge but disseminate patriotic and humane values. The expectations were that the intelligentsia would be the embodiment of high morals and values, national heritage, etc. Almantas Samalavicius describes how it sustained the national culture in Lithuania:

> Despite of enforced indoctrination and vulgarization of discourses (...) humanities in Lithuania and other Baltic countries succeeded in performing many roles and were largely instrumental in shaping national consciousness and collective historical memory that survived throughout half a century of foreign rule. (...) The end of dependence was, however, marked with the rise of excesses of national feelings, exalted patriotic rhetoric and politicized discourses – factors that for a few years

> delayed the formation of truthful visions and adequate understanding of these countries' economic, social and cultural realities. (Samalavicius 2003: 52)

Now, after fifteen years of transformation, other problems appeared, instead of nationalism, the excessive commercialization of any sphere of public life including higher education. The Intelligentsia is transforming itself into a group of professionals – technically but not morally skilled experts in various fields. This issue seems to be a big dilemma for any such group in Western civilization: What is an 'appropriate' development for universities to aspire to and what role should intellectuals play? Probably the major problem is that students are more and more rarely taught critical thinking or encouraged to ask difficult questions. They are rather taught and they rather want to be taught how to acquire a diploma in order to lead a comfortable and conformist life.

Eric Gould published a whole book (*The University in a Corporate Culture*) devoted to this issue which seems to be really crucial for the further existence of Western civilization. The books reviewer summarizes it as follows:

> What place does a liberal education have in a society where getting a good job after graduation is the dominant force behind a college degree? Should universities change their focus from producing independent, free-thinking individuals to turning out ready-made workers? Are college students today less interested in what they're learning than students of past generations? (...) Students today are more likely to think of a college degree purely as a passport to a job or career (Christner 2003: 99).

In Poland during the post-communist system transformation and European integration the introduction of a decentralized, plural political system and a free market economy affected most of the so called 'budget sectors' including education. The most obvious transformations were visible in academic institutions and the tertiary level of education and their commercialization was both the most painful and at the same time the most 'successful'. Many schools now started to concentrate on quantity rather than quality.

First of all the whole system of higher education was changed. After 1990 non-state higher schools were introduced as well as higher vocational state schools. The diversification of the system has of course advantages and disadvantages. A decentralized system may be difficult to control – so a drop in the quality of education may take place, but on the other hand particular sections may prove to be more flexible. One of the most visible and evident features of this metamorphosis is how the numbers increased and thus the quality decreased. In Poland during the last ten years the number of students was increased by three times (current number of students – 1.8 million people) and the number of higher education institutions grew by four times (there are about 400 state & non-state higher schools: http://studia.korba.pl/szkola_przetrwania/).

Table 1 presents the growth of the number of schools and students, and only a slight growth of the number of academic teachers.

	1995/96	2000/01	2002/03	2003/2004
Higher schools	179	310	377	400
Students in thousands	794.6	1584.8	1800.5	1858.7

Table 1 (Source: *Concise Statistical Yearbook of Poland 2004*)

Maybe the academic staff will be enlarged by graduates who more and more often choose Ph.D. studies as their next stage in career (see Table 2.)

Year	Number of Ph.D. scholarships
1990	1868
1991	1490
1992	1644
1993	1921
1994	2412
1995	5393
1996	7753
1997	9481
1998	11359
1999	12364

Table 2 (Source: http://www.buwiwm.edu.pl/publ/przew)

The research on Polish students, especially in association with the issue of the intelligentsia, is quite a popular topic in the time of transformation. The reasons are connected with the dimensions of time and space: A few major research projects were carried out demonstrating that Polish students believe in the future of Poland – a new Poland as a member of the European Union. The transformation seems to be the variable responsible for the growing objectivity among students (com. Szymborska-Kubiak 2003). Students are also analysed in the context of socio-cultural capital and market aspirations of a future Polish intelligentsia (com. Bartoszek 2003). Academic culture, too constitutes an interesting topic associated with the aspects of agency in students especially if Polish students are compared with their counterparts in other nations (e.g. students in Germany, com. Gieseke, Półtorzycki 2001). Lately Polish students have been analysed in the context of labour force and labour market in research reports published by AIG OFE and Gazeta Wyborcza (www.gazeta.pl/mlodzi 2005).

The portrait of an average Polish student during the period of transformation emerging from these studies is composed of a few features and qualities: more and more reflexive, more and more often connecting studies with work, more and more eager to learn foreign languages, and study two or more faculties, more and more often studying abroad.

Preliminary Research

My idea about this preliminary study project was first of all to ask students more questions about their plans for the future as combined with this special moment of our nation's existence. I wanted to check whether reflexivity and agency of the individual are somehow connected with the reflexivity of the collective and to see whether individual biographies are complimentary with the 'national biography'. Poland has made a huge effort to join the EU, it was our state's priority and major collective drive. We were informed each day by the mass media about the completion of further stages of the process which was maybe one of the main tasks and targets in our history. It was started mainly by our elites (formerly known as intelligentsia) but slowly almost all classes, groups and categories of individuals were incorporated, involved in the project. Provided we make the assumption that students are our future elites it would be interesting to find out what projects they are going to fulfill as this may lead us to some clues about our national or even European future.

I asked the respondents questions on their plans to leave the country, on their attitudes towards their peers leaving Poland, about whether they consider themselves as patriots in order to find out whether our future elites will share any qualities with the former intelligentsia which was the carrier of national values, strict moral virtues and devotion to the motherland and its history.

Some specific results

In Table 3 the options for the question 'what do studies mean to you?' are presented (respondents were given 8 options and they were asked to tick as many as they wanted). The most common associations with studies are: self-fulfillment, an interesting job in the future, better knowledge of the world and the self and contact with interesting people.

Option	*Number of students who selected the option*
More interesting job in the future	167
More prestige	125
Contact with interesting people	161
Waiting for better economic situation	32
Longer careless youth	83
Satisfying aspirations, self-fulfillment	180
Better cognition of myself and the world	165
Perspective of earning fortunes	57

Table 3 (Source: compiled by the author, preliminary research)

In the next question respondents were asked to imagine what their lives would look like in 10 years time and to indicate the most important factors in such an outlined biography. They had to put the options in order from the most to the least important. The most significant life project priorities for young Poles taking part in this survey are: to acquire a good education, to have a job, to have a family/ a partner (Table 4).

Options	Selected in the first place	Selected as second	Selected as third
I will be well educated	84	36	28
I will have a job	31	75	34
I will have a family	32	19	32
I will own a flat	15	12	50
I will have a child	5	12	16
I will experience interesting travels all over the world	13	17	14
I will have a successful career	17	22	19

Table 4 (Source: compiled by the author, preliminary research)

Further questions concerned the issue of having short- and long-term goals and whether respondents would accept their friends' decision about leaving Poland for a permanent life abroad. In questions referring to goals the respondents could choose the options: yes, no, sometimes. For question about their friends' decision they could tick: definitely yes, rather yes, difficult to say, rather not, definitely not. (In Table 5 below not all the options are presented). The majority of respondents have long term goals and most of them accept a friends' decision to leave Poland. Proportionally there are more people against their friends' decision to leave Poland among respondents with no long term goals.

Acceptance of friends' decision on leaving Poland	Long term goals *yes*	Long term goals *no*
Definitely yes	41	15
Rather yes	37	10
Rather not	2	3
Definitely not	1	0

Table 5 (Source: compiled by the author, preliminary study)

Table 6 presents a similar juxtaposition referring to short term goals. Very few people among the respondents have no short term goals.

Acceptance of friends' decision on leaving Poland	Short term goals *yes*	Short term goals *no*
Definitely yes	57	6
Rather yes	65	4
Rather not	6	0
Definitely not	3	0

Table 6 (Source: compiled by the author, preliminary study)

In further questions students were asked whether they perceive themselves as patriots and whether they have the conviction that they have an influence on their lives. (In questions about patriotism the respondents had the following options to choose from: yes, rather yes, difficult to say, rather not, not; in questions referring to the influence on one's life: yes, not, to some extent).

In Table 7 the question about a patriotic declaration is confronted with the sex variable. In the research sample there is no gender correlation with or lack of patriotism. 142 people out of 200 answered positively. Men chose the option of lack of patriotism slightly more frequently.

Patriotism	Total	Female	Male
Definitely yes	*49*	*22*	*27*
Rather yes	*93*	*50*	*43*
Total positive answers	142	72	70
Rather not	*15*	*5*	*10*
Definitely not	*11*	*5*	*6*
Total negative answers	26	10	16

Table 7 (Source: compiled by the author, preliminary study)

In Table 8 answers for the questions about patriotism and about the influence on one's life (agency) are juxtaposed. 91 people out of 142 who positively voiced their patriotic leaning are convinced that it definitely has an influence on their lives. Among the people who think of themselves as non-patriots there are proportionally fewer people convinced of their agency than among the ones who answered positively to the patriotism question.

Patriotism	Influence on one's life - *yes*	Influence on one's life - *no*	Influence on one's life – *to some extent*
Definitely yes	35	2	12
Rather yes	56	4	33
Total positive answers	91	6	45
Rather not	5	2	8
Definitely not	6	1	4
Total negative answers	11	3	12

Table 8 (Source: compiled by the author, preliminary study)

But what does patriotism mean nowadays? The next table shows that it does not have to mean staying in the motherland for ever. The majority of the sampled respondents think of themselves as patriotic but still the majority wants to leave Poland at least for some time.

Further questions were designed to find information whether those who believe they have an influence on their lives want to leave Poland? In this preliminary research sample more men than women are willing to leave for a few years. The other way round – more women than men want to leave for good (Does this reflect women's sense of a longer perspective or men's more patriotic nature?) But answers of rather not and definitely not were also more frequent among women. Respondents were also asked to predict the Polish intelligentsia's role in the European Union. The answers may be divided into three basic categories: negative connotations with the intelligentsia (particular interests, making money), the intelligentsia is perceived as a group of experts thanks to whom the EU will benefit and Poland will enjoy good representation, no idea (and no responses).

Plans to leave	Have influence on their lives	Don't have influence on their lives	Have some influence on their lives
Yes, for good	*10*	*2*	*2*
Yes, for a few years	*28*	*4*	*23*
Yes for a few months	*30*	*1*	*27*
Total positive	68	7	52
Rather not	*17*	*2*	*16*
Definitely not	*8*	*1*	*3*
Total negative	25	3	19

Table 9 (Source: compiled by the author, preliminary study)

One may draw the conclusion that former associations with the intelligentsia are not familiar to the young generation and that future elites (intellectuals) will resemble the Western European knowledge class – a loose group of professionals with no aspirations to be the leader for the rest of citizens in terms of defining future directions of development, judging what is right and wrong both for individual and society.

Conclusion

Now, although the majority of young students think of themselves as patriots, they are willing to go abroad for at least a few years if not longer. They accept the decisions of their peers to leave and are quite convinced about having an influence on their lives, about their agency. The majority of them think about their future in terms of short-term goals to achieve and most of them also work out long-term goals. Students as a future elite will be less bound to the motherland although they will still be thinking in terms of national interest (patriotism). They will be much more mobile and stricter about planning and monitoring their lives.

Definitely further research is needed as this was not a representative sample. Furthermore a larger research project would be more suitable to answer more profound questions. Will increasing reflexivity result in improving the social system or just upgrade the chances of young people to make a quick career? Will reflexivity make current students reinterpret future capitalism and help to transform it into a friendlier environment for the previously underprivileged or will it just be a tool for making big money rather than a tool for shaping personalities searching to provide their lives with more meaning and sense? The question still remains open.

Works Cited

Bartoszek, Adam (2003). *Kapitał społeczno-kulturowy młodej inteligencji wobec wymogów rynku*. Katowice: Wydawnictwo Uniwersytetu Śląskiego.
Beck, Ulrich (2002). *Społeczeństwo ryzyka*. Warszawa: Wydawnictwo Naukowe Scholar.
Christner, Terry (2003). "The University In a Corporate Culture (Book review)". *Library Journal* 128.12, 99.
Concise Statistical Yearbok of Poland 2004 (2004). Warsaw.
Cremin, Colin S. (2005). "Profiling the Personal: Configuration of Teenage Biographies to Employment Norms". *Sociology* 39.2, 315.
Devine, Nesta & Ruth Irwin (2005). "Autonomy, Agency and Education: He tangata, he tangata, he tangata" *Educational Philosophy and Theory* 37.3, 317.
Giddens, Anthony (1991). *Modernity and Self-Identity. Self and Society in the Late Modern Age*. Cambridge: Polity Press.
Giddens, Anthony (1984) *The Constitution of Society*. Cambridge: Polity Press.
Giddens, Anthony & Christopher Pierson (1998). *Conversations with Anthony Giddens. Making Sense of Modernity*. Cambridge: Polity Press.
Gieseke, Wilhud & Józef Półturzycki, ed. (2001). *Kultura studiowania w okresie transformacji*. Warszawa: ŻAK Wydawnictwo Akademickie.
Hofstede, Geert (1991). *Cultures and Organisations. Software of the Mind. Intercultural Co-operation and its Importance for Survival*. London: Harper Collins Business.
Koralewicz, Jadwiga & Marek Ziółkowski (2003). *Mentalność Polaków*. Warszawa: Scholar.
Kubiak-Szymborska, Ewa (2003). *Podmiotowość młodzieży akademickiej*. Bydgoszcz: Wydawnictwo Akademii Bydgoskiej.
Marshall, Gordon, ed. (1998). *A Dictionary of Sociology*. Oxford: OUP.
McAuliffe, Garrett J. (2002). "Student Changes, Program Influences, and Adult Development in One Program of Councelor Training: An Exploratory Inductive Inquiry". *Journal of Adult Development* 9.3, 205.
McLeod, Julie (2002). "Working out Intimacy: young people and friendship in an age of reflexivity". *Discourse: Studies in the Cultural Politics of Education* 23.2, 211.
McLeod, Julie (2003). "Why we interview now – reflexivity and perspective in a longitudinal study". *International Journal of Social Research Methodology* 6.3, 201.
Meyer, John W., Jepperson Ronald L. (2000). "The Actors of Modern Society: The Cultural Construction of Social Agency". *Sociological Theory* 18.1, 100.
Nelson, Laird Thomas F. (2005). "'College Students' Experiences with Diversity and Their Effects on Academic Self-Confidence, Social Agency, and Disposition toward Critical Thinking". *Research in Higher Education* 46.4, 365.
Samalavicius, Almantas (2003). "Humanities and Academic Cultures: Towards Dialogue". *Dialogue and Universalism* 1-2, 51-58.

Catarina Frois

Identification and Anonymity

Two Sides of the Same Coin*

Who are you?
I really want to know…
(The Who)

The lyrics of the track that play over the credits at the beginning of an episode of the TV serial C.S.I. – Crime Scene Investigation – poses the crucial question around which this essay revolves: "Who are you?". It is a categorical way of illustrating how this becomes a recurring issue we have to deal with in a variety of everyday life situations that involve dealing with others. There would seem to be a contradiction or opposition between the name we answer to or the different technologies used to identify, recognize and distinguish us, and our attempts to escape those same devices resorting to disguises, false identities, pseudonyms or anonymity. But in fact they turn out to be parallel processes. The need for identification and the escape through anonymity go hand in hand. There are many reasons that can explain the variety of actions that individuals assume socially. For the moment, it is important that we think about the conditions that have made these two phenomena so pressing.

I will strive to put in contrast the following basic ideas: the need for identification against the need for anonymity. As I have said, these phenomenons are not considered here as independent, I think of them as parallel, ways of interaction and communication within society; an important part of an individual's reality as a social being. Interdependency in this case means, in our view, that new technology for individual identification is accompanied by new forms of attaining anonymity, both processes simultaneously validating each others strategies.

The first part of this essay follows the book *Documenting Individual Identity*, edited by Jane Caplan and John Torpey (2001). It is used here mainly

* I wish to thank Gary T. Marx, David Lyon and Charles Raab for their comments to earlier drafts of this paper.

as a reference that may help us understand the modernization of identification procedures in an ever more complex society.

The second part will take a closer look at anonymity, focusing especially on the way it is experienced in those self-help groups denominated "anonymous". The data used for analysis was almost entirely researched for my PhD in Anthropology during fieldwork carried out between October 2002 and March 2005 with Families Anonymous, Alcoholics Anonymous and Narcotics Anonymous in the Lisbon area, Portugal. This part of the paper will discuss the way anonymity proves to be fundamental in the course of the action, assuming specific and apparently contradictory features. Here, I will also approach the idea of the possibility of anonymity in a face to face context.

I am aware that the scope of these issues cannot be contained in a short paper like this one, and that some aspects may be left unexplored or partially so. I will not go into great detail on the issue regarding the distinction between identification and identity; the idea of surveillance and all that it entails, will be considered mainly as an important reference; the same will be the case with the idea of privacy. I will nevertheless try to include these issues, acknowledging the interest they hold for scholars in this area and related areas, since it crosses multidisciplinary concerns in an advanced theoretical and ethnographic context.

Identification as an Issue

As much as has been written on *identity*[1] in the social sciences, there has been little reflection around *identification*, or the role played by identification in the way an individual defines him/herself. Even considering that the different processes of identification are external to the individual – in the sense that what we are dealing with here specifically is not the subject presenting him/herself, but the variety of mechanisms that allow us to know him/her – we admit that the vast array of documentation, of physical evidence that can be used to identify a singular body, tell us nothing about who we are as subjects with a personal history and life experience. They are, in effect a fundamental part of our movements and everyday experience, of our attire, of the everyday objects we carry around; all of which are essential in the relationships we maintain with others. Poster states: "Social security cards, drivers' licenses, credit cards, library cards and the like – the individual must apply for them, have them ready at all times, use them continuously" (1990:93).

Whereas in the Middle Ages recognition was made mainly through the use of symbols, insignias or attire, as in the case of travelers for example, in the second half of the 18th century, we see the beginning of a concern to create records with information on individuals on a large scale, using the name, marital

[1] See Giddens (1991), Stuart Hall (1996), Herzfeld (1997).

status, place of birth or address. We witness the birth of bureaucracy and the bureaucratic State in all its extension. From the moment that the concern with identification of individuals starts, it is connected with keeping written records, which need qualified personnel who will not be tied up by personal motivations, but who will simply enforce the law equally for all. This constitutes an essential element in the definition of bureaucracy in its widest sense. Following Max Weber's idea

> the growing complexity and specialization in modern cultures demand for its external structure an employee who will be personally uninvolved and strictly objective, to substitute the lord of earlier times, who was moved by personal feelings, favors, sympathies and gratitude (1976: 48)[2].

On one hand, we have the growth of population in the cities, the resulting increase of mobility and anonymity; and on the other hand, the need for a bureaucratic machine that will treat all subjects on equal terms, making no distinctions of any kind, and can keep records containing information regarding each individual.

The ways to identify a person extend in different directions as we move forward in time and consider different cultural and historical settings. Later on, passports are introduced (internal and external) to somehow control the movements of citizens within and beyond the frontiers of their countries of origin. Other technologies have been developed in this process: fingerprint records, anthropometric measurements, and the attempt to establish physical patterns that may somehow allow 'predicting' deviant tendencies or propensity for relapse; decoding human DNA; surveillance by closed circuit television cameras; the use of the body as a source of visual identification through the deciphering of the elements that make it unique.

The appearance of these techniques permits us to contemplate issues that transcend the function for which they were initially used. Therefore, if at first they were intended primarily for the identification of criminals and to figure out possible genetic patterns that might lead to deviant behavior, they quickly started being used as a means at the disposal of official utilities on the whole, irrespective of ends for which they were meant. Some of the objections raised by this use are related mainly to the fact that this transformation started to be shared indiscriminately. Fingerprint records were initially only taken for criminals or in crime-related situations; in the case of DNA samples, their first use was chiefly medical. In both cases, once their utility was proved as a way to identify and

---

[2] About bureaucracy, surveillance and modernity see Dandeker (1990), Herzfeld (1992), Poster (1995, 1990).

differentiate all and every single person, they were quickly adopted, and became considered a part of our bureaucratic routines.

What I consider relevant in this example is the 'adaptation' of individual identification techniques and technologies, for purposes other than those initially considered. This is all the more evident when we think about the goals involved: crime prevention, the control of individuals to ensure their safety, commercial-oriented, and consumer profiling purposes, etc. In other words, new/old instruments of identification are met by new/old needs. Nelkin & Lindee, in their study on the cultural appropriations of DNA, also claim:

> Despite continued controversy over methods and motives, efforts to determine the genetic basis of human behaviors such as alcoholism and crime draw legitimacy from the rising fortunes of molecular biology. These efforts have captured public attention, for such research addresses critical social questions – about the basis of human identity and individual differences, the nature of deviance, and the local of responsibility for social problems (2004: 10).

In relation to surveillance, I am reminded of the concept of the *Panopticon* mentioned by Michel Foucault. The idea behind this device was the total monitoring of an individual by another within a prison, creating to that end facilities that would make it possible. This concept is widely spread today through the use of closed-circuit surveillance cameras (CCTV) that we find everywhere to a greater or lesser degree, according to the context in question (see Marx, 1995). It may be worth to go over Foucault's idea so that we can later appreciate, by comparison, the currency of his thought:

> The Panopticon is a device intended to dissociate the pair see/be seen [...] to be totally seen without ever seeing [...] an important device since it de-individualizes power [...] so it becomes irrelevant to know who exerts it. Almost anybody, taken at random, can operate such a machine (1996: 167).

As Norris & Armstrong observe, however, despite the differences as to the purpose of the surveillance mechanism described by Foucault, when we compare it with the uses of CCTV (in the case of the Panopticon, Foucault describes its function as the indoctrination and conditioning of the behavior of the individual being punished; whereas with CCTV the uses and purposes are more varied although it doesn't track a person from one location to the other, i.e., different cameras in different places are not connected in a way that allow the total monitoring of an individual), I am led to admit that the principle involved is equal for both. A totalitarian device with regard to the individual which results in his/her movements being followed, recorded and, wherever it may seem adequate, punished. As in the passage quoted previously, Norris and Armstrong also admit the advantages of a de-individualized surveillance device – that is, not influenced by its operator's will – but, when they consider the use of machines as instruments, they take this idea to the limit:

> Human visual capability has difficulty competing with the high surveillance capabilities of the camera: the camera does not blink, sleep or get bored and, unlike images captured on videotape, the results of human visual surveillance cannot be rewound or replayed in a court of law (op. cit.: 18).

Another issue that deserves attention in this frame of analysis is raised by David Lyon (2001) and has to do with the fact that a person's body, as a crucial element for the subject's identification, is increasingly devoid of individuality. If at a basic level, the choice of name for a person becomes a central feature of the way the individual is presented to others, and contains a personal history and experience that is by itself an *individualizing* feature, we have to admit that nowadays knowing (in the sense of legal identification recognition) who this or that person is, doesn't necessary imply a knowledge of his/her personal, emotional or cultural history, but exclusively of his/her body. As the author puts it:

> Once it was merely the existence of unique bodies that was part of the rationale for individuation, and for stabilizing difference. But now, for example, through fingerprint, other signs of bodily distinctiveness are appealed to. Direct checking from the 1970s on became a matter of verification by a third-party organization. This was done digitally by methods such as data matching since dataveillance regimes were electronically established in the 1980s. But from the 1990s, it became clear that direct checking would take in yet another meaning: access to tissues, fluids, images, and patterns available from the body itself. Just as direct checking across agencies avoided confrontation with the embodied agent, so direct checking of data produced from within bodies also requires no access to the speech or the memory of the person. It is, once again, abstracted from the person (2001: 296-297).

Here, we are in a position to rethink the old debate within social studies between the idea of individual and person[3], in light of the new elements for analysis raised by the issues described. If we consider Lyon's proposition, it seems legitimate to add a new element of analysis to this equation: the body of the individual or person, regardless of the other features described above. In other words, the anonymous body, identical to all other bodies yet at the same time unique and distinguishable. The body as an amorphous entity, that can be objectified independently of who is inhabiting it. We are no longer considering the person or the individual's will to be different, but the fact that it may be so, regardless of the subject's will; that is, the body as a separate element.

---

[3] See Carrithers, Collins & Lukes (1985), Fortes (1987).

In this first part, my goal was to cross the problem of identification with other directly related issues: bureaucracy, modernity, urbanity, the body, individual and person, privacy or surveillance. I found that processes of identification have become increasingly complex with time, keeping up with various social transformations. This issue is not merely a question of the 'legal' or 'official' identification, but also of the need individuals have to identify themselves and others by means of social and moral control. In my view, it is the awareness of this need that triggers the discussion of control-freedom, which is translated in this essay into identification-anonymity. As there are different motives that contribute to our need to know the identity of those interacting with us, this can mean that it assumes different features. Therefore, while we find that there is a wide range of ways to identify us, at the same time we realize that the simultaneous need to act 'incognito', to remain anonymous, and as such, unidentified, is also related with the factors described previously.

I now propose a brief analysis of the notion of anonymity in the social sciences so that we can next discuss this concept in view of the empirical information collected during fieldwork with associations that make anonymity a rule.

Anonymity in Social Sciences – A Brief Account

> Une foule d'inconnus s'y croisent constamment. Il est une manière d'être citadin, inculquée dès l'enfance, qui consiste à marcher dans la rue sans sauter comme un cabri ni montrer les passants du doigt, sans se faire remarquer, individu semblable aux autres, neutres et anonymes. L'anonymat est au coeur du phénomène urbain. Il règne en maître dans les lieux publics, protecteur de chacun, du soi non révélé, condition aussi nécessaire que la précédente à la coexistence de millions d'habitants (Pétonnet, 1987: 249).

Anonymity's importance as an issue for reflection can be quite simply traced in broad terms. It has become increasingly relevant in the field of social studies. Furthermore, and despite the growing concern of this issue for sociologists, anthropologists, historians and other scholars, I realize that this issue suffers constant change and that there is still room for exploration. I propose a broad (and necessarily short) approach to the way in which this phenomenon has been dealt with recently. To mention but a few examples, we discover anonymity being analyzed in relation to issues connected to the use of Internet and information society, literature, phenomenology, privacy, anthropology.

Helen Nissenbaum (1999, 1998) takes the idea of anonymity in its wider sense, as it is defined in the dictionary (the absence of a name for the subject of an action) to argue that in a society dominated by information technologies and by the recognition of the other, anonymity is not simply the absence of a name, but the possibility of remaining "unreachable". The sociologist Gary Marx

(1999) is thinking along these lines when he lays down a conceptual framework, combining anonymity and identification, in an attempt to chart the kinds of situations in which the appeal to anonymity is justified, as well as the need to identify subjects. Jo Ann Oravec (2003) discusses the way anonymity and privacy assume ever more diverging forms, by the combination of multiple contexts in which these phenomena can occur, or be at risk; arguing that they constitute identity management devices for individuals.

Robert Griffin (2003) reflects on the use of anonymity and/or pseudonyms in literary works (from the 16th to the 20th century), arguing that anonymity is like a mask that authors choose to put on at the moment of publication – regardless of their motivations for choosing to do so – and which is in itself an identifying mark of the author. Maurice Natanson (1979), following Alfred Schutz's thoughts regarding the "anonymous", considers nomination, or the act of nominating, as being at the core of anonymity, and of the relationships established among individuals in everyday face to face situations. Daniel Terrole (1996), in his work on the homeless ("Sans-Domicile-Fixe") considers anonymity to be part of a liminal process, in which the subject gradually looses his/her personal distinctive features. Milandou (1997), based on the case study of the relationships between natives of Brazzaville, reflects on the confrontation between the anonymity which is typical of big cities, and the strategies employed by their inhabitants to be conspicuous, at least around their own neighborhood, in an attempt to preserve ways of interaction that are typical of smaller communities.

What all these views have in common is the fact that they consider anonymity to be related with the absence of some kind of identity element of the acting subject. As we have seen at the beginning of this essay, these elements are a crucial part of our daily routines and of the way we interact with others. We might take this to be a definition of anonymity in its wider sense, which in our view is consensual as to: 1) the way anonymity is defined in the dictionary (absence of the name of the subject in an action); 2) the different contexts and case studies presented here. However, I will now introduce some elements that will help us to divide this concept further, considering anonymity to be not only a form of personal information management, but in fact a condition that must be obeyed in some contexts for interaction to exist. The elements which I will present next, allow us to formulate these proposals based on the fact that I consider anonymity to be a way of managing the personal information we give others and a necessity that must be complied with so that the action may successfully unfold in the physical presence of others.

Anonymity in Anonymous Groups

The data presented herein are the result of a *work in progress* since 2002, initially with Families Anonymous, and later with Alcoholics Anonymous and Narcotics Anonymous. Fieldwork was carried out with groups chosen in the Lisbon area (Portugal), in participant observation of weekly meetings, at different locations and schedules. In addition, interviews were held with group members, trying, whenever possible, to register different stages of membership.

Alcoholics Anonymous (AA) focus on problems related to alcohol abuse. In my experience, the groups in this association are made up mainly of female members in an age group ranging between 35 and 60 years of age. Narcotics Anonymous (NA) deal with issues concerning "addiction".[4] In other words, they are concerned with drug and alcohol abuse, or abuse of other chemical substances. Contrary to the previous group, I observed a predominance of male members between 25 and 45 years of age. Families Anonymous (FA) is meant for people who have a relative with an addiction problem, mainly women between 45 and 65. All these associations are exclusively run by members, without the intervention of professional specialists in the different areas, thus their definition as self-help, or mutual-help groups.

As a result of sharing a common model, all three associations describe themselves as Fellowships, Non-profit Organizations, with a particular statute and organization, in which certain individuals meet to "solve" the problems that brought them there through a process of mutual identification and the sharing of common feelings and situations. According to their philosophy, the *12 Steps* are "stages" or "principles" that the subject must go through, in order to gradually reach a stable situation, both physically as well as emotionally, regarding the problem they are dealing with. Along with the *12 Traditions* (rules that dictate the way the group behaves), they are the basis for these associations.

Meetings take place in diverse locations, although we registered a preference for religious buildings (churches and parishes), and an attempt to avoid "exposed" places, so as to ensure the privacy of anyone who seeks them. In other words, the choice of site depends on availability of rooms (and the agreement of a landlord) rather than on other kinds of issues. This fact becomes even more interesting if we compare it, for instance, with the existence of surveillance cameras in the cities, as I mentioned earlier. Even though in Portugal this technology is not widely spread (it can only be found in major cities), when it comes to choosing a space to hold meetings in, there is a

---

[4] The term addiction is used here in the sense commonly given by the informants, without further discussion of the concept. On this subject, see Davies (1997). There are other works focusing on therapeutic groups that follow the 12 Step model, such as Keane (2000), Ronel (2000), Cain (1997), Antze (1987).

preference for discrete sites, which can not be immediately connected with their activity and are protected from public scrutiny. In the context of these anonymous groups, members value above all else the absence of a moral judgment on the part of outsiders concerning their addiction as drug addicts (NA), alcoholics (AA) or relatives of people with these problems (FA).

The members of anonymous groups want to be able to participate and do not want their acceptance to depend on the elements that are socially significant outside the group. We might say, therefore, that from the moment they step into a meeting, or when they speak as a member, wherever this may be, that they strip off all other attributes and become simply individuals with a specific problem that is common to everyone and constitutes their only bond. Elements of identity– such as their place of residence, profession, education or marital status – are destitute of their importance in this context.

All of this can only be properly assessed if we understand, on one hand, the emotional charge that lies behind the motives that drive a person who seeks this kind of association, and on the other hand, the issue of anonymity. In fact, the existence and protection of anonymity are considered crucial for all the action to unfold in this manner. It is one of the basic principles in the philosophy of all these associations. Anonymity is discussed here, not simply in its most common sense, which has to do with the absence of a name or any other form of individual identification, but equally important, as a form of protection searched by individuals in this context.

As we have shown in the first part of this paper and considering that in contemporary society it is increasingly more difficult to pass unrecognized, we realize that this fact does not, however, reduce the need individuals have of going incognito in certain circumstances in the course of their lives. If, as we have found, identification is only partial – since it only reveals the legal definition of a person and omits elements of one's social, cultural and emotional individuality – in the case of anonymous groups, it is precisely this 'flip side of the coin', so to speak, that is revealed. It is precisely the particular configuration of the context in which the action occurs, that allows only certain elements of an individual's identity to be revealed, namely as a subject with a certain condition, given that anonymity is also only partial and limited.

In these associations anonymity is respected outside the group as well as within it, despite face to face contact occurring in a meeting. Initially, 'not having to say who you are' proves to be paramount, even if, as Erving Goffman (1983) claims, all face to face interaction, including non verbal interaction, implies knowledge of the other. In this particular case, the fact that members come together to discuss a specific problem (common to all) without having to reveal everyday situations that could expose them – such as who they are, where they live – is determinant, and renders all other descriptions irrelevant.

Anonymity is considered a guarantee and a condition of inclusion of the members in these associations, at the same time this is considered a therapeutic vehicle in itself. It applies both inside and outside the group. This means that anonymity is respected within these groups as to who and what is revealed there.

Outside the meetings, anonymity is used to guarantee the confidentiality of what is said by whom, and even the membership of subject *a* or *b*. This results not just from the stigma that people feel, but also from the fact that they find anonymity to be a leveling element, abolishing any differences that they have between them beyond the specific motivations that drew them there.

> *When you are there, it doesn't matter if you are a lawyer, or a doctor; if you are wealthy, come from royalty or off the street. It doesn't matter, that is not important, the name is not important, all those things don't matter. The only thing on your mind is your intention of starting something new in which drugs have no part* (Diogo, 38 years old, has been in Narcotics Anonymous for 5 years).

Regardless of the way relationships develop among individuals in this context – we find that initially they are marked by some embarrassment and discomfort, but later assume different features – we must not rule out anonymity in a face to face interaction. In other words, even if at a certain stage, members start developing bonds of friendship and trust that go beyond the limits or goals of group meetings, the important thing is the choice provided by this rule. The freedom to choose what to say, to whom and when, is in itself a way of making therapy possible, as well as the groups and meetings themselves.

Conclusion

In the course of this essay I tried to discuss two issues that, while different, are interdependent: the need to be identified on one hand, and the need to remain anonymous on the other. I believe that both these needs cannot be truly understood separately, since we can only understand the real proportions of either when we consider their origin, and only realize their relevance if we measure one and the other.

My goal was to account for different perspectives surrounding the issue of identification and anonymity. In the first part of this paper, I tried to show that underlying this notion there are many others that can be looked at separately, as for instance urban growth in the modern age; the creation of egalitarian and uniform forms of identification that keep up with people's increasing mobility; crime-fighting policies; the creation of surveillance systems; the need to profile consumers, markets and trends, etc. In short, we have a whole range of mechanisms that allow us to know more about people's actions, whereabouts and habits, in a world where technology and science play an ever growing role in the way we relate to each other.

Identification and anonymity can therefore co-exist, and the point is to distinguish between the different types of identification – legal, emotional, physical – and the type of anonymity we are dealing with, considering that anonymity is partial, relative and limited in space and time. In the case of anonymous groups, I tried to describe the role played by anonymity, not just as a rule, but as a necessary element which becomes the *leitmotiv* that makes relationships possible, effective and specific. Anonymity becomes a choice that people can use for themselves, for the identity they want to communicate, whether they are physically present or otherwise. Individuals in this situation are concerned, above all else, with their motives, and it is based on them that they manage their identity, and that ultimately answers the question we opened this paper with: "Who are you?" Anonymity is a mask the member puts on, both within and without the group, in their contact with an "outside world" that they feel to be judging. In these groups, anonymity is more than the possibility of managing personal information; it becomes a common referent of interaction. It is a choice and an option that determines the whole action. I have tried to show that there is more to anonymity than the difficulty in attaining it in an environment of sophisticated detection and identification technology. Based on the data collected during fieldwork, I arrive at two kinds of conclusions: individuals find ways to 'escape' surveillance (in this case study it is chiefly moral); and, in the physical presence of another, an individual adapts himself and his personal information and directs it to a specific goal depending on the situation. Anonymity is therefore an alternative to physical, moral and social control and is constantly adapting to circumstances.

Works Cited

Antze, Paul (2003 [1987]). "Symbolic Action in Alcoholics Anonymous". *Constructive Drinking*. Ed. Mary Douglas. London: Routledge, 149-181.

Cain, Carole (1991). "Personal Stories: Identity Acquisition and Self-Understanding in Alcoholics Anonymous". *Ethos* 19.2, 210-253.

Caplan, Jane & John Torpey, ed. (2001). *Documenting Individual Identity*. Princeton: Princeton UP.

Carrithers, Michael, Steven Collins & Steven Lukes, ed. (1985). *The Category of the Person*. Cambridge: Cambridge UP.

Davies, John Booth (1997). *Drugspeak: The Analysis of Drug Discourse*. Amsterdam B. V.: Harwood Academic Publishers.

Dandeker, Christopher (1990). *Surveillance, Power and Modernity*. Cambridge: Polity Press.

Fortes, Meyer (1987). *Religion, Morality and the Person: Essays on Tallensi Religion*. Ed. Jack Goody. Cambridge: Cambridge UP.

Foucault, Michel (1977 [1975]). *Discipline and Punish*. Tr. Alan Sheridan. New York: Vintage.

Giddens, Anthony (1991). *Modernity and Self-Identity: Self and Society in the Late Modern Age*. Stanford: Stanford UP.
Goffman, Erving (1963). *Stigma: Notes on the Management of Spoiled Identity*. New Jersey: Prentice-Hall.
Griffin, Robert J., ed. (2003). *The Faces of Anonymity: Anonymous and Pseudonymous Publications from the Sixteenth to the Twentieth Century*. New York: Palgrave Macmillan.
Hall, Stuart (1996). "Who Needs Identity?" *Questions of Cultural Identity*. Ed. Stuart Hall & Paul du Gay. London: Sage, 1-17.
Herzfeld, Michael (1997). *Cultural Intimacy: Social Poetics in the Nation-State*. New York: Routledge.
Herzfeld, Michael (1992). *The Social Production of Indifference*. Chicago: Berg.
Keane, Helen (2000). "Setting Yourself Free: Techniques of Recovery". *Health* 4.3, 324-346.
Lyon, David (2001). "Under My Skin: From Identification Papers to Body Surveillance". *Documenting Individual Identity*. Ed. Jane Caplan & John Torpey. Princeton: Princeton UP, 291-310.
Marx, Gary (1999). "What's in a Name? Some Reflections on the Sociology of Anonymity". *The Information Society* 15, 99-112.
Marx, Gary (1995). "Undercover in Comparative Perspective: Some Implications for Knowledge and Social Research". *Undercover: Police Surveillance in Contemporary Perspective*. Ed. Cyrille Fijinaut & Gary T. Marx. Norwell: Kluwer Academic Publishers, 322-337.
Milandou, Augustin-Marie (1997). "«Type connu! Qui le connaît pas?» Anonymat et culture populaire à Brazzaville". *L'Homme* 141, 119-130.
Natanson, Maurice (1979). "Phenomenology, Anonymity and Alienation". *New Literary History* 10.3, 533-546.
Nelkin, Dorothy & M. Susan Lindee (2004). *The DNA Mystique: The Gene as a Cultural Icon*. Ann Arbor: U of Michigan Press.
Nissenbaum, Helen (1999). "The Meaning of Anonymity in an Information Age". *The Information Society* 15, 141-144.
Nissenbaum, Helen (1998). "Protecting Privacy in an Information Age: The Problem of Privacy in Public". *Law and Philosophy* 17, 559-596.
Norris, Clive & Gary Armstrong (1999). *The Maximum Surveillance Society: The Rise of CCTV*. Oxford: Berg.
Oravec, Jo Ann (2003). "The Transformation of Privacy and Anonymity: Beyond the 'Right to Be Let Alone'". *Sociological Imagination* 39.1, 3-23.
Pétonnet, Colette (1987). "L'anonymat ou la pellicule protectrice". *Le Temps de La Réflexion VIII: La ville inquiète*. Paris: Gallimard.
Poster, Mark (1995). *The Second Media Age*. Cambridge: Polity Press.
Poster, Mark (1990). *The Mode of Information: Poststructuralism and Social Context*. Cambridge: Polity Press.
Ronel, Natti (2000). "From Self-Help to Professional Care: An Enhanced Application of the 12-Step Program". *The Journal of Applied Behavioural Science* 36.1, 108-122.
Terrole, Daniel (1996). "Anonymisation et défense collective". *Ethnologie française* XXVI. 3, 418-425.
Touraine, Alan et al. (1976). *A Burocracia*. Lisboa: Sociocultur.

Steven Corcoran

Postulating Equality

Politics and Philosophy in the Oeuvre of Jacques Rancière

Questions of thought, for Jacques Rancière, come under issues of combat, of relations of force. His interventions on political questions are always well circumscribed, agitated, but cautious in edging toward the inescapable shores of politics. Interventionist more than meditative, the temporality of his philosophical practice harks to the imperative of an action, the thinking deployed in the political event.

What exactly is the place of philosophy for Jacques Rancière in relation to the practice of politics? What style of critical practice does it yield? And what are the strengths and possible limitations of this critical practice? I shall attempt to answer these questions in three parts. First I shall sketch Rancière's conceptualisation of politics. Then, I shall show how Rancière's practice of philosophy relates to his conceptualisation of politics. Finally, in concluding, I will briefly touch on a seeming difficulty with Rancière's approach.

Politics and egalitarian subjectivity

A brief detour through the Rancière's unique response to the question of political subjectivity and the specific difference it introduces into the social field will prove crucial to localising his specific practice of philosophy.

So, first, an elementary but necessary question: What is politics for Rancière? How does a political subject come to pass? Rancière distinguishes politics from every exercise of power or simple organisation of the affairs of a pre-existing community. It cannot be deduced from the essence of language or the activity of communication. Politics rather has to do with "the coming to surface of the latent force of the dominated," a demonstration in which this latent force is revealed as the generally "hijacked source of the machination" of the powers that be (Badiou 2005: 123).

As Rancière's reading of Aristotle makes clear, every property (e.g. the possession of language) from which one tries to deduce the politicalness of humankind is always presented as divided. The possession of language is not a common property enabling man to engage in "politics," understood as the activity of leading and commanding. Rather, in every social order, this "possession" is always divided, divided between those who really "possess" speech and those who only "understand" it; those whose qualities and titles to rule is "founded" on a class of others who are relegated to the merely private or social sphere.[1] But in order for the distribution of places to function, those who are told merely to obey – to occupy their places, to inhabit their social identities – must first be addressed as equals, i.e. as beings that make sense, understand language and recognise in general. Politics then emerges when those excluded from the common argue not only that their voices be heard on an equal footing with the ruling oligarchy or aristocracy, that is, that their projects also be taken into account etc. It involves an aesthetic reconfiguring of space by an impossible subject, by a subject that emerges on the sole basis of the empty equality that founds the social order, and whose declaration takes the following form: we, the no-part, are the people, the stand-in for All, because we are the only ones that testify to the principle of equality, over and against all figures of private interest.

Otherwise put, politics takes the form of an essentially *dissensual* struggle between two operations: between the hierarchical social structure (or "police": the order of counting in which every part is assigned a place and a quality of speech) and a "non-part" which suspends relations of domination on the basis of the universal maxim of the equality of anyone and everyone. The non-part subjectivates its lack of place, takes the time to demonstrate its equality with those who rule, the effects of which force a suspension of the "natural" functional order of social relations by introducing a difference – the difference of equality – into the partition of the perceptible governing the distribution of places and functions.

As Rancière thus shows against Habermas, politics is not a rational debate between multiple, relevant interests. The Habermasian schema presupposes that the objects and partners of interlocution are pre-established, whereas politics involves precisely a struggle over who is counted as a legitimate speaking partner and what objects can be debated. To gain a voice in the shaping of common affairs, those without part must put in common that which is not given as common (i.e. the quality of their speech). But since the non-part has no title to assert, it can only argue on the basis of a paradoxical title, namely, the title of those without title, or the simple equality of everyone and anyone. Its most elementary gesture involves this short-circuit between the universal and the particular, whereby those excluded from the common, demonstrate this wrongful exclusion on the basis that their position is the universal position. Every

[1] See in particular chapters 2 and 3 of *Disagreement: politics and philosophy.* (Rancière 1999).

political subject invents *cases of universality*, cases which when taken into account rupture every count legitimating the rule of the few. Every egalitarian politics thus involves an attempt to speak on the common whose condition is a refusal to speak from an assigned place. Politics is therefore in no sense the result of individuals or groups, identified by the interests they come together to debate. It always involves a de-objectification, or *declassification*, of speech, and can always be situated in relation to a specific police situation of injustice. Politics polarises a particular situation but it does not do so in some Schmittian formal life and death struggle with the enemy. Politics has its own measure, which is equality, and the opponents of politics here are not substantial others threatening one's way of life. The political adversary is one who proclaims that the natural order (of domination) is all there is, one who declares himself to be against the political gesture of "impossible" equality. For, despite the universal measure of political argumentation, it will always provoke reaction by those to whom consensus on the merely possible enables rule by wealth.

Exploiting an ambiguity in the Marxist schema, Rancière therefore argues that class struggle proper lies not between two identifiable classes, one of whom has a universal vocation, but between the non-part and those who oppose its claims to equality. Workers in the nineteenth century, for example, struggled valiantly not simply to obtain a condescending recognition of their importance in the production process but to construct the world of work as a common object and not simply as an object belonging to the world of private relations and interests. Elaborating on the Marxist intuition, Rancière argues that politics involves not merely a debate over interests but a disagreement over the separation of spheres, over the right to construct different common objects in which a more egalitarian world is possible.

The paradox of politics, of a part of those without part that suspends the count of the oligarchic police, implies a radical reorganisation of the relation of philosophy to politics. Political philosophy purports to be the science of the *polis*, a programmatic enterprise that seeks to found or reflexively think the essence of the political community. It takes politics to be a universal experience that it has the task of thinking in its own register. Better still, it supposes that there exists a sole organising principle, *the* political, from which the community, its laws and contests for power all hark in truth. But the presupposition of an *arkhè* – an underlying principle – assumes that there are reasons behind the linking of certain individuals to certain "political" functions. It presupposes, in other words, that there are *reasons* for the social distribution. Who does not see that this implies a vicious circle? If the political relation is deduced from a mode of life that, posited as the mode of life proper to politics, has the universal or the good as a specific element, then politics gets founded on a mode of life destined for politics![2]

But the paradoxical demonstration of the title to rule of those who have no title has the precise effect of doing away with the hierarchisation of modes of life and the privatisation of the common by the custodians of state. Every genuine politics has the effect of *un-linking* speech from assigned places. As such, politics never simply involves the assumption of power by the dominated since this would merely reverse the relation between dominated and dominating, keeping intact the separation of spheres. Instead, it is a matter of introducing cases of universality and capacities of formulating the common into what was considered the social, private or domestic universe. The notion, dear to political philosophy, that there exists a pure sphere of political purity is what egalitarian politics constantly puts into question. Rancière thus rigorously shows that, by supposing the division on which it bases its "science," political philosophy ends up presupposing as the foundation of politics the division that constitutes its object.

What both the elementary circle of political philosophy and the tautology of the police distribution of the social (obey because you must!) share is the attempt to eliminate the litigious argumentation of equality. What one seeks to foreclose by the power of its count, and the other by its circularity, is the latent power proper to the politics of equality and the non-localisable nature of political conflict. Indeed, Rancière argues, this exclusion of the masses from thought is *constitutive* of political philosophy itself as the very endeavour that undertakes *to do politics in truth*. The scandal at the heart of political philosophy, the paradox that brought it into being, visible to the authors of its inaugural texts, is that politics be "founded" on the paradox of a part of no-part.[3]

A new practice of philosophy

At the most general level, it might be said that the relation between the politics of philosophers and real politics is one of constant, unfolding combat. For Rancière, the task of the philosopher is to render visible what belongs to each in this struggle where the two sides are forever intermingling, forever pillaging the arsenal of the other in order to perfect themselves. The important thing to see here is that politics is conceptualised as a figure of intellectuality in its own right. And it happens regardless, or in spite, of political philosophical attempts

[2] Cf. "Dix thèses sur la politique."

[3] Cf. *Dis-agreement*. This is, according to Rancière, the paradoxical virtue of Plato. He recognises this title but then rejects it. Aristotle, for his part, claims it cannot be rejected but ends up reducing it by appropriating its elements.

to occult the radicality of dissension from political debate and project a figure in which a "true" politics would finally "right" democratic chaos.

Otherwise put: politics is thus is no way an *object* of philosophy, in the traditional sense. Philosophy is not that which brings its legislation to the law or its reflection to the ends of the community. Instead, what is specific to "political philosophy" lies in the way it relates to the paradox of politics by seeking to reduce it, to reject, appropriate or displace it.[4] Insofar as it exists, it exists in relation to the very paradox, or conflict, that it attempts to eliminate. It exists, that is, by its *effets en retour*: the effects it has of blurring or covering up of the distinction between police and politics, the sole condition on which consensus can reign as to the localisation of politics in the state. But at the same time, the mobility of politics undoes the pretensions of political philosophy – renders it inexistent or vain – and makes another kind of philosophical practice necessary.

We can already glimpse here that such a practice is for Rancière the capacity to separate "real politics" – i.e. that which happens – from the "politics of philosophers" – the principled melancholia of the absence of a true politics. Philosophy here is not a matter of commentary nor a question of the cognitive appropriation of an object (the political). Philosophy is an operation of separation. It traces a line between the forms and rationality proper to politics and that proper to the police, on the one hand, and the philosophical attempts to effectuate a cover-up, on the other.

But if philosophy cannot be identified by a preconstituted field (i.e. an object), this is not because, as with Althusser, philosophy has no object and finally no relation other than to itself. It is rather that philosophy has always singular objects, is "made of nodes of thought," as Rancière puts it. Where philosophy meets politics, art, science or any other domain in which thought is exercised, it is under the sign of a specific paradox, conflict, or aporia. Philosophy, therefore, contrary to the vast ongoing practice of "political philosophy," does not qualify as political because it is called upon to lend its reflection or its legislation to a great form of human activity. "Philosophy becomes "political" when it embraces aporia or the quandary proper to politics." Rancière argues that this was true even for Aristotle himself:

> [He] points this out in a phrase that is one of the first encounters between the noun *philosophy* and the adjective *political:* 'Equality or inequality, comes down to aporia and political philosophy' (Rancière 1999: ix).

Rancière thus affirms that only when thought embraces aporia is philosophy born and effects of thought produced.

---

[4] Rancière names the figures of rationality that attempt to reduce politics (Platonic) archi-politics, (Aristotelian) para-politics, and (Marxist) meta-politics.

He thus distinguishes himself from the interior of philosophy from every phenomenological or hermeneutical conception. Philosophy is neither a form of epistemological apprehension, nor a form of questioning. It is an act of declaration. Thus, if Rancière takes extensive pains to detect and reactivate the sediments of politics under the layers of discourse, this is not with a phenomenological view to discovering meaning, a primordial ground. What Rancière finds, buried beneath strata of discourse, is precisely another discourse, one traced in the *après-coup* of an event bearing the logic of disagreement. What philosophy, as an act of declaration, declares is nothing other than the paradox of the rationality of politics itself, which it promptly welcomes as a proper philosophical object. If "the *mésentente* of politics by philosophy has as its principle the very reduction of the rationality of disagreement," (Rancière 1999: xii) if this defines the basic operation of all the great texts of political philosophy from Plato and Aristotle, through to Hobbes, Marx (!) and Arendt, then Rancière's style of philosophy distinguishes itself precisely with regard to the way it welcomes the quandary of politics into philosophy.

Welcoming the quandary: this obviously cannot imply the "pious vision" according to which "philosophy comes to the rescue of the practitioner of politics [...] explaining the reason for his quandary by shedding light on the principle of his practice." (Rancière 1999: ix) For if this were the case, politics would return to the position of an object whose principle it would be the task of the philosopher to explicate; whereas Rancière is at pains to establish that the rationality of politics is *anarchic* and that, as a figure of intellectuality in its own right, it does not call upon the philosopher to come to its rescue.

The difference might be elucidated by showing in what way the philosophy-effect is here different to the knowledge-effect produced in the sciences. For the localising of the event precisely undoes the pretension to hierarchy and specificity of the various knowledges and is essential to one of the main functions of Rancière's work: the discrediting of postures of governmental and intellectual mastery. Rancière never tires of showing that every mastery is an imposture in that it supposes representations of which the montage is fallacious. That they are fallacious is established with reference to the "local expressions of the non-mastery of the dominated," (Badiou 2005: 123) which pull the rug out from under the master.

The philosophy-effect here stems precisely from a redoubling of the power of the event and its names (demos, proletariat, workers, women, and so on). If there is no object properly speaking but only a subjective declaration and localisation of equality, then philosophy for Rancière is simply the place that redoubles that power. The act of philosophy is a declaration, a declarative intervention that traces a line in the theoretical field with respect to the historical externality that is politics. Philosophy delimits, separates. It is the activity that separates out movements of thinking within theoretical disciplines according to their possible effects of class, in Rancière's sense of that term.

It is here that we may locate the rupture that Rancière introduces into philosophy with respect to contemporary avatars of political philosophy. It is not only that it refuses every attempt to interpret the world, to declare on this or that basis what the state of the world is and what should be done to remedy it. It is that it is the only philosophy not to deny or reduce the specific node where philosophy, from time to time, meets politics. It is the only philosophy not to deny the effects of equality in the domain of theory, and to unleash them. It is the only philosophy to draw the radically egalitarian effects of political invention for theory, and to demonstrate the effects of division, of dissensus, that interiorising the presupposition of equality implies.

Or, rather, Rancière's practice of philosophy is the only one not to disavow its possible effects of class and to control its effects of division within philosophy and the entire theoretical domain. We could say that it is a philosophy that is in an enlightened position with regard to the paradoxical object of egalitarian practice that mediates its position. This enlightened position is not one that has a better medicine for correcting of practice. Since we have said that philosophy cannot be identified as political on the basis of an existing field or object, it cannot therefore know how it effects its political condition. Rather, this enlightened position is one that, by interiorising the egalitarian maxim, controls the effects of exclusion stemming from theoretical postures of mastery. It is one that undermines everything in theoretical constructions that purport to separate the scientist's domain from the field of everyday experience, and hence its perceived use for or effects on a politics of emancipation. As such, it is a philosophy for which these possible effects, ever an unknown supposition, are different.

Rancière's practice of philosophy then goes hand in hand with his idea of politics: it is anarchical. Anarchical: that means he subscribes to the basic presupposition of the equality of all intelligences. The equality of all intelligences does not mean that every manifestation of human intelligence is the same, patently not. It means "that the same intelligence makes poetic fictions, political inventions, or historical explanations, that the same intelligence makes and understands sentences in general." He continues,

> Political thought, history, sociology and so on, all use common powers of linguistic innovation in order to make their objects visible and create connections between them (Rancière 2000: 116).

This means that all discourses, including philosophy, occupy the same plane; that all activities where human thought is exercised refer to a fundamental equality. Philosophy, however, seems to occupy a special place within the multiplicity of practices – though obviously not as a discipline that grounds other forms of discourse or spheres of rationality. Instead, it is the discourse that works to undo the boundaries that separate disciplines and grant them their authority on the supposed basis that there are specific methodologies that fit specific fields of objectivity.

To this end, Rancière deploys what he calls a poetics of knowledge, which is the theoretical counterpart to an aesthetics of politics. We have said that the act of philosophy is the invention of a tracing that redoubles, within itself, the effectiveness of the politics of equality to distinguish it from the pretence of the politics of philosophers. But it is slightly misleading to portray this operation as strictly immanent to philosophical practice since it operates precisely in the interval of discourses, between philosophy, sociology, political science, history and so on: wherever impostures of mastery provide forms of description of the world that extend the effects of political philosophy.[5] For Rancière, the office of philosophy is to circulate between these discourses and the political supplement, without taking them as objects (i.e. separate fields of which the mechanism can be divulged) but by submitting them to a poetics of knowledge. Philosophy's task is to trace back the specificity of disciplines and discursive competencies to the egalitarian level of linguistic competence and poetic invention. To do this, Rancière *subtracts* the texts of political sociology, of history etc., from all "the forms of self-legitimation upon which they rest by arguing for the specificity of their objects and methods" (Rancière 2000: 116). He demonstrates that these discourses all utilise common powers of linguistic innovation to make objects visible and available to thinking and to create connections between objects. This requires breaking with our most entrenched intellectual habits and the invention of a way of

> reintegrating these discourses into a generally accessible mode of reasoning or form of language so that everyone may partake in this creative activity of invention that allows for a redescription and reconfiguration of a common world of experience (Rancière 2000: 116).

The complimentary procedure involves *extracting* speech acts and political texts from political sociology to give them a positive value as such. In *La nuit des prolétaires,* for example, Rancière *extracts*, the texts of workers and other political declarations, usually treated as mere expressions of a condition or ideological artefacts, from their socioeconomic links and situates them on the same level as other texts. He is thereby able to elide a sociocultural reading and look at such texts or performances as inventions of forms of language similar to all others. For, Rancière argues, it is only in the revendication of the efficacy of

[5] Cf. Thesis nine of "Dix thèses sur la politique" in *Aux bords du politique,* p. 247.

the literary that the political valence of mere speech acts, of the egalitarian powers of language, indifferent with respect to the speaker" can be put into play (Rancière 2000: 116).

As Rancière himself puts it:

> An object of reflection commands the aperture of a specific temporality. That is, it commands the presence of a process of writing, or the construction of a specific form of writing, oriented toward an intrusive encounter with a specific mode of thinking that, in its turn, creates a particular thought-event by interrupting the organisation of a class of objects or a series of performances. Thinking for me is always a rethinking. Thinking means to submit an object of thought to a specific variation that includes a shift in its discursive register, its universe of reference, or its temporal designation. [...] It is this staging that determines the conditions for a constitutive rethinking; that is to say, it is a restaging. The elaboration of these 'moments of thinking' is for me the task of a philosophy that challenges the boundaries separating the classes of discourses (Rancière 2000: 120).

The theme of the inventive tracing of discourses to their common powers is thus the consequence of the singularising of the object. For if philosophy, placed under the condition of politics, is nonetheless not able to treat politics, it can only support it by the invention of a *process of writing* specific and immanent to that which conditions it. Philosophy does not do the thinking of politics. Instead it thinks the power of certain *names* (the proletariat, emancipation) and invents forms of writing to render that politicalness. Its tracing from that moment onward only has effects in itself, by which it is modified.

We know that the transgression of politics, *in rupture* with the social order and the present of the police, creates a dissensus in the present of relations of command. Interiorising that rupture means creating a rupture, a dissensus, with certain authoritative theoretical constructs. Rancière again:

> To construct a specific present...thus requires a double transgression. On the one hand, it is incumbent to transgress the divisions of discourse: divisions that separate the disciplines (philosophy, political science, history etc.), the divisions of noble and profane discourse, the divisions between a logic comprised of links in a chain of real events and the logic of a chain of fictional events. On the other hand, it is imperative to revoke the authoritative principle derived from the succession of historical events. And it is the implications derived from this second transgressive imperative that I understand to be critical to an idea of contemporaneousness. To conceptualise the contemporaneousness of thought requires a reliance on a certain anachronism or untimeliness (Rancière 2000: 115-16).

Which means that, as neither a theory of practice, nor a history of the sciences, philosophy is practical but nevertheless ahistorical.

To recapitulate: philosophy, for Rancière, cannot be identified in relation to a preconstituted field and ultimately has no relation except to singular objects.

That is to say, its real effect is entirely produced within the space of thinking instituted by the paradoxes and conflicts arising between the political events and its intellectual and governmental disavowal. But as we saw above, the supposition of Rancière's work is that it is precisely through the effects it produces in itself that philosophy intervenes in the real.

Now, if philosophy intervenes in reality through the immanent operation of welcoming singular nodes of politics as philosophical objects, then wherein lies the real of those effects? This boils down to knowing how philosophy affects its singular objects, and, more precisely, how philosophy returns on its politics if it is not in the form of epistemological apprehension. The style here cannot be to oppose, in the tradition of political philosophy, a politics of truth to politics as it is generally considered (the inconsistent exercise of power), since that would mean that thinking politics would once more appear in the form of thinking an object (the political) separated from the social or the private. The clear consequence of the "singularisation of the object" is that philosophy cannot know how it affects politics.

However, this does not mean that philosophy (or metapolitics) has no external effects i.e. effects on the practice of politics in relation to which it operates. Because it is one accidental activity among others that can be situated in the general field of practices it has external effects. But philosophy has such effects, that is, intervenes in reality, precisely through the effects it produces in itself. Philosophy can act on external practices by the result of the verification within itself of the egalitarian axiom. But philosophy can never know the effects it has on politics. As Rancière puts it in an interview, "All I can define are forms of perception, forms of utterance. As to how these are then taken up by organisations" is not a matter of theory (Rancière 2003: 200). Philosophy does not prescribe rules for the just practice of the egalitarian axiom i.e. for militant practice. Rather, it can only hope, in defining forms of perception, for an opening of space: once put into circulation who knows how philosophy might aid the reinvention of politics today.

It is in *Disagreement* that Rancière makes the most systematic attempt to define these forms of perception. With great erudition, Rancière separates out the forms of perception that belong to politics from those that belong to the police. When, today, the power of words like "emancipation," "workers" and so on, has all but evaporated, Rancière deftly demonstrates that this is not because the old social divisions have simply been superseded, yielding a more fluid, indeterminate and consensual flux of identities all inhabiting a single field of force relations. Rancière knows that politics is not about the representation of social groups or identities, and never has been. If this statist view of politics is so prevalent today it is simply to do with the receding and forgetting of the forms of democratic politics that current modes of governance and their intellectual justification at once appropriate and reduce. To put it another way: the indistinction today between the parliamentary left and right is not simply due to the complexification of society and the multiplication of identities. As Marx knew, it is in the absence of political intervention that society comes to seem evasively complex, a network of multiple and overlapping classifications. And perhaps the clearest symptom of this transformation of the democratic struggle is the status of the *foreigner:* no longer able to inscribe him- or herself in the space of litigation to protest against injustice as a member of the working class, the "foreigner" is conferred a status that is never properly "regulated" and appears as the foreign body come to steal our work.

Though philosophy cannot tell what effect the defining of forms of perception may have on politics, it is clearly of important *strategic* value. For, nothing in politics itself is capable of indicating the ultimate importance of an event in the shaping of public space whenever it comes to pass. There is nothing in politics that indicates itself as politics from the outside. It falls to philosophy to analyse political sequences and, moreover, the forms of reaction or resignation that follow in its wake. Politics happens regardless of philosophy; but it is crucial for politics, perhaps especially in times of its eclipse, to find its philosopher.

The relation of primacy, however, is clear: first politics and then philosophy. Nonetheless, as we've seen, philosophy retains an essential autonomy with relation to politics. This autonomy consists precisely in the invention of scenes of dissensus by which it becomes possible to discern the effectiveness specific to politics in the theoretical domain. And, on this basis, it allows for a fully inventive critical practice, one that, by referring the statements of dominant discourses and representative government to the effectiveness of the political supplement, can construct scenes that discredit (im)postures of mastery.

Concluding remarks

This recasting of the enterprise of grasping politics, and of the critical power it authorises, is not a thinking which offers a new solution to the problem of the foundation of the community or the legitimacy of government. Rancière does not seek to tell us what, in an uncertain future, we must do to govern society politically. It is a thinking of the political event as such, of history as rupture, of politics as the suspension of the exercise of power and its system of legitimation. Philosophy that is political is thus a revolution in excess of the revolution, one that redoubles the power of the event. More precisely, it is an *intervention* that endeavours to break the circle of political philosophy and cognate discourses through an egalitarian flattening out of all discourse.

It should thus be clear what, in times of the absence or crisis of egalitarian politics, the strategy of this interventionist philosophy is: to recall the imperative of politics and the ancient struggle between politics and philosophers. In the current conjuncture, this means for Rancière to distinguish a type of research into the present crisis of democratic contestation from two apparently opposed discourses: that which proclaims the end of encroachments on the political by the social and hails a return to politics; and that which laments that the time of 'grand' politics has passed and declares that humanity's destiny is no longer bound up with the theme of emancipation but with warding off catastrophe. Rancière introduces a difference into the lamentable present portrayed by these would be master narratives by reminding us of the precariousness of politics. Politics has neither returned in its purity nor durably passed with the overcoming of the "illusion" of emancipation. Rancière has little trouble showing that these twin discourses, premised on the forgetting of the egalitarian event, are inconsistent. But his approach cannot be straightforwardly to denounce these discourses and to deliver principles for the practice of a properly egalitarian politics i.e. one that could realise equality. At most, what Rancière does is to lay paradoxes, ones occasioned by the forgetting of politics, or its absorption into the social, across the line of premature conclusions, so as to halt them in their tracks, and reopen the space for the perception of the consequences of a possible future postulation of equality.

Indeed, Rancière is at his most persuasive in demonstrating that the possibility of a new politics cannot be derived either from identitarian fundamentalism or from the experience of the inhuman. Neither identity politics nor the injunction to attend to radical otherness could help us found a new politics: they are two sides of the same coin, at least to the extent that they are both premised on the specific foreclosure of egalitarian politics. However, is it the case, as far as theory is concerned, that the only conclusion to be drawn, our only consolation, rests with the idea of the precariousness of politics?

Rancière will tell us what politics has been, what it is no longer, and how the consensual police would have it dissolved in the self-representation of identities or lost in the desert of humanitarian expansion. But his critical practice will not tell us what should be argued, or what should be done so that its polemical equality is inscribed in the public space. The project of political philosophy, which is to ground a true politics *against* the egalitarian effects of the people, may well be a vain enterprise. Does this mean that philosophy is limited to a simple retroactive nomination of egalitarian politics?

Rancière demonstrates that, if we want to think politics, it is indispensable to carry this out with regard to the words and actions of those engaged in politics by which the relations of domination are, albeit inconclusively, suspended. As an inventive redoubling of the power of the event, his critical practice is a formidable tool.[6] Yet in refusing every ontological approach to theorising the

[6] For more recent developments see his latest book *La haine de la démocratie*. Paris: La fabrique, 2005.

event, Rancière can only conceptualise politics as an instance that takes as its object the politicisation of something previously considered social or private. Reciprocally, in the absence of politics, Rancière can show how its elements have been, to borrow a term from Gilles Deleuze, reterritorialised in a new police order. Rancière can tell us what politics has been, what it no longer is, and what it is not. But he is unable to suggest what it might be today or what we should do to make it exist here and now. Rancière's historical "phenomenology" of the egalitarian instance limits itself to the modesty of grasping effects of equality and the power of the names attached to them. His anti-ontological approach manifestly prohibits every attempt to grasp politics as a singularity, in the singularity of its being, or in the purity of its form. The result of course is that Rancière is unable to propose any normative solution to the impasse of post-politics, or to affirm any prescription or mode of organisation so that politics might exist. That this reticence is shared by political philosophy suggests, however, that we need forms of philosophical intervention capable of doing just that.

Works Cited

Badiou, Alain (2005 [1998]). *Handbook of metapolitics.* Tr. Jason Barker. London: Verso.

Rancière, Jacques (1998) "Dix thèses sur la politique". *Aux bords du politique.* Paris: Gallimard.

Rancière, Jacques (1999 [1995]). *Disagreement: politics and philosophy.* Tr. Julie Rose. Minneapolis: University of Minnesota Press.

Rancière, Jacques (2000) "Dissenting Words: A Conversation with Jacques Rancière". *Diacritics* 30.2, 113-26.

Rancière, Jacques (2003) "Politics and aesthetics an interview". *Angelaki: the One or the Other? French Philosophy Today (special issue)* 8.2, 191-211.

aesthetic representations

Dirk Wiemann

Mountaineering with Hobbes and Milton

(R)Enunciations of Seeing and Agency in Restoration England

After the 1660 restoration of the monarchy in England and the concomitant rollback, the unresolved question of political enfranchisement and agency retains the potential explosiveness that had become manifest in the form and course of the English Revolution itself. Mere containment can at most make it go underground, where it continues to express itself in various disguises that still employ the 'dangerous' vocabulary of the revolutionary period. If, as has been amply documented, the register of (radical) Protestantism provided the master code for the discursive struggles fought out in the intensely mobilised public sphere of the Civil War and Revolution (see, eg, Jameson: 84; Holstun; Hill), it needs to be added that in this conflictual field of political/religious discourses a rhetoric of seeing attains a particular centrality. This becomes most apparent in the manifold expressions of popular politics that, in the course of the Revolution, tend to get encoded in prophetic and/or apocalyptic terms of 'discovery' and revelatory 'disclosure'[1]. This way of seeing survives the Revolution itself most prominently in Bunyan's *Pilgrim's Progress* (1678): a text introduced with the claim to be "delivered under the Similitude of a Dream Wherein is Discovered The manner of his setting out" (Bunyan 29), and shot through with iterative interpolations in which the narrator asserts that the entire narrative consists of what "I saw in my dream". Yet, though framed as a divinely induced vision, Bunyan's text has to accommodate an altogether different way of seeing in its folds: At one of the stations of their quest, the pilgrims are led by a group of helpful shepherds "to the top of an high Hill called Clear" (Bunyan 161). From this height, the Celestial City – the final destination of the pilgrimage and the narrative – can be spotted in a proleptic vision that is also, strikingly, prosthetic. For it is only with the help of the shepherds' "perspective glass" that "something like the Gate, and some of the

[1] This applies to the popular left as well as the popular right spectrum: From Winstanley to Thomas Edwards and the *Eikon Basilike*, visionary insight and inspiration form a current topic.

glory of the place" can be perceived from the elevated vantage point of Mount Clear. The pilgrims' view of the beyond, in short, is surprisingly 'modern': Not prophetic but scientific, it is performed in the manner of contemporary land survey[2], dependent on the skilled usage of technology[3] rather than on 'inspiration'. Moreover, the very name with which the optical prosthesis is introduced as the 'perspective glass' underscores how the religious vision is articulated with the modern visual regime of linear perspective and the functional space it produces.

Such attempts to articulate the religious master code with the early modern perspectival turn (and the ensuing rearrangement of socially produced and represented space) can be pursued further in relation to the struggle over agency. In the following, I am going to read two paradigmatic Restoration texts – Hobbes's *Behemoth* (publ. 1682), and Milton's *Paradise Lost* (1674) – as widely discrepant interventions into the contest over agency with particular emphasis on the politics of the specular at the interface of Renaissance 'anthropocracy' and theocratic interdiction. Both texts thematise vision (more precisely: panoramic vistas from mountain tops) as a privileged metonymy of agency – fiercely delegitimised by Hobbes, emphatically endorsed by Milton. Obviously, then, these bird's-eye-view fantasies are implicated in a visual regime that reconstructs space and subjectivity in a specifically 'modern' way.

Seeing in perspective

One of the most striking novelties of the early modern visual regime appears to lie in the introduction and naturalisation of a virtual, non-physiological way of 'seeing', for which the "artificial perspective" as elaborated by Alberti, Leonardo, and Dürer may serve here as a shorthand.

> The effect of this invention was nothing less than to convince an entire civilization that it possessed an infallible method of representation, a system for the automatic and mechanical production of truths about the material and mental worlds (Mitchell 37).

According to Erwin Panofsky's classical iconology, the invention of perspective in 15th-century Italian painting[4] ushers in the "structure of an infinite, continuous and homogeneous – in short: a purely mathematical – space" (Panofsky 101). In

---

[2] Bernhard Klein depicts the early modern "land surveyor [...] climbing the highest mountain of the region, or demanding admittance to the local church steeple best situated for taking accurate measurements of the surrounding landscape" (Klein 3).

[3] For the introduction and refinement of optical measurement equipment, see Tyacke 2001.

[4] Perspectival painting in Britain has its heyday only after the decline of Elizabethan pseudo-medievalism, i.e. in the first half of the 17th century; see Wilton 26-51.

this genuinely modern construction of 'rational' space, objects and bodies are primarily conceived of as relational "points" on a three-dimensional grid. Inasmuch as they are determined by their specific location instead of their substance, modern bodies – as Ernst Cassirer observes – do no longer possess "any independent content of their own. Their existence is reduced to their reciprocal relationalities: it is a purely functional, not a substantial being" (Cassirer 107). Other than in all preceding spatial constructions, then, modernity posits objective space in a striking analogy to Walter Benjamin's famous dissection of homogeneous empty time (Benjamin 260—63): that is, as continuous, homogeneous, infinite and essentially *empty* inasmuch as space is now figured as always already there prior to, and independent of, the emergence of any body located within it.

Its systematic objectivity notwithstanding, rational Renaissance space is yet simultaneously "dependent on man, on the individual" (Panofsky 123) on whose eye the perspectival artefact is rigorously centred. Accordingly, for Panofsky, perspective functions as the formal expression of modern anthropocracy as it emerges from the vestiges of medieval theocracy and, more precisely, from the "virulence of a deficit in the foundational structure of medieval knowledge" (Schmeiser 47). John Berger, pushing Panofsky's argument to its extreme, observes: "Perspective makes the single eye the centre of the world. Everything converges on to the single eye as to the vanishing point of infinity. The visible world is arranged for the spectator as the universe was once thought to be arranged for God" (Berger 18).

If Renaissance painting thus locates the spectator at a "deified position" (Holly 198) and inaugurates an oxymoronic "objectification of the subjective" (Panofsky 123), it would be worth the while to retrieve the ambivalence of that term, 'the subjective'. For as both Panofsky and Berger emphasise, the perspectival painting posits the spectator as a figure that is itself a mathematical reduction: Perspective does not aim at a mimetic reproduction of the physiological spheroid field of vision but simulates instead the gaze of a single, fixed eye. The fact that the *trompe l'oeuil* aspirations of Renaissance and mannerist painting do all the same find their realisation is hence owed, not to some strict verisimilitude, some coincidence with the "actual structure of the subjective visual impression" (Panofsky 104), but much rather to the implementation, and naturalisation, of linear perspective as the crucial component of the modern visual regime – in other words, the reordering of the culturally dominant way of seeing: The spatial order of modernity has to suppress both the curvatures and the heterogeneity of antique and medieval constructions of space in order to achieve both the centrality of the spectator (at the expense of his stereoscopic vision) and the continuity and homogeneity of space itself – a rational space into which the individual is interpellated as a desubstantialised, 'purely functional' body even while being elevated to the status of centrality. Panofsky mentions in passing that Kepler still testifies to the curved lines of comets that later astronomers tended to describe as straight trajectories; and he goes on to speculate that, "if hardly anybody living today has ever seen those curvatures [mentioned by Kepler], then this is certainly owed to our habituation to the perspectival construction" (Panofsky 104).

Seeing thus occurs at the interface of subjectivity and subjection. The subject's entrance into the visible world is always an entrance into a discursively prestructured field, into which the seeing eye may be sutured as anthropocratic centre to such an extent that anthropocracy itself finds its primary manifestation in the act of seeing; all this, however, can only occur on the condition that seeing structurally abide by the rules of perspective and thus collaborate in the production and consolidation of modernity's empty, homogeneous space. The ambivalence of seeing in early modernity is in that sense nothing but the ambivalence of agency itself: a subjectivity that can only be constituted by interpellation on which it ineluctably depends but of which it is also potentially in excess. Like Benjamin's homogeneous empty time, so is the rational and functional space of early modern projections ultimately fictitious[5], always flexed under the pressure of 'tradition', and in need to be "accommodated to a pre-existent world view" (Morgan 55).

Political pornography: With Thomas Hobbes on the Devil's Mountain

The Hobbesian model of the state appears to be intimately tied in with the fiction of a continuous and homogeneous perspectival space in which bodies are reduced to 'merely functional points'. Most obviously, the architecture of the absolutist commonwealth according to *Leviathan* rests not on some transcendental essence, some substance inherent in the ruler, but on the immanent, purely contractual and representational relation between the sovereign and his 'authors':

> in him consisteth the Essence of the Commonwealth; which (to define it,) is One Person, of whose Acts a great Multitude, by mutuall Covenants one with another, have made themselves every one the Author, to the end he may use the strength and means of them all, as he shall think expedient, for their Peace and Common Defence (Hobbes 1996: 121).

What Hobbes blows up to the status of a founding moment of good governance consists in the general transfer of 'natural rights' from all individuals onto the artificial figure of the absolute monarch. While emphasising the authority of the

[5] See Partha Chatterjee's critique of Benedict Anderson's employment of Benjamin's notion of homogeneous empty time; Chatterjee 161-170.

multitude in the establishment of the contract, Hobbes makes it clear that the covenant is designed to assign a monopoly on agency to the ruler as "actor". Once the contract is sealed (and it is always already sealed), agency – the right to 'act' – rests with the absolute sovereign alone. Hobbes's alertness to the histrionic implications involved here[6] renders the fictitious moment of general transfer readable as a displaced continuation of the spectacular performance of absolutism so popular at the Tudor and Stuart courts before the English Revolution: the masque. This highly stylised form of enactment involved royalty and nobility as 'actors' but more precisely as perspectival points in rational space: The masque is as much an installation as it is a play, invariably culminating in a *tableau vivant* in which the king, seated on his royal chair, is encircled by his courtiers who all focus their looks on him. The king, then, forms the still centre of a circle intersected by centripetally arranged perspectival rays reminiscent of spokes in a wheel. While certainly this theatre of looks intends to dramatise the gaze as a manifestation of precisely that transfer, or discharge, of authority that Hobbes has in mind, it all the same reasserts the agency of the bearer of the gaze. The masque, therefore, permanently points to the fact that "the monarch is always a monarch *with courtiers*" (and by extension, subjects in general): "The royal figure is frozen into a representation of sheer 'power', but this power can be exercised only through the courtiers" (Sinfield 84); more than that, the sovereign's prerogative to agency is, in a performative contradiction, denied to the monarch as "the lines of perspective that focus upon the king's chair [...] not only privilege the king's position, they also hold him trapped" (Sinfield 84). In short, the visual theatre of the masque, relying on the artificial Renaissance perspective with its unprecedented centralisation of focus, has to simultaneously acknowledge that this perspective is always ineluctably centred on the individual spectator; it thus, despite its intent to enact monarchical supremacy, has to confirm the specular agency of the individual and hence decentralise the visual field. In order to contain these ambivalent effects of perspectival vision, Hobbes, I will argue, has to make seeing itself the monopoly of the legitimate actor(s). For this, he takes recourse to precisely that religious register that his genealogy of the State purports to supersede. This becomes most apparent in the opening paragraph to *Behemoth*, the post-restoration account of the parliamentarian 'sedition' of the 1640s and 1650s:

> If in time, as in place, there were degrees of high and low, I verily believe that the highest of time would be that which passed between the years of 1640 and 1660. For

---

[6] See Hobbes 1996: 112: "So that a *Person*, is the same that an *Actor* is, both on the stage and in common Conversation; and to *Personate* is to *Act*, or *Represent* himselfe, or an other; and he that acteth another, is said to beare his Person, or act in his name".

> he that thence, as from the Devil's Mountain, should have looked upon the world and observed the actions of men, especially in England, might have had a prospect of all kinds of injustice, and of all kinds of folly, that the world could afford [...] (Hobbes 1990: 1; subseq. *B* + page number).

Behemoth is thus introduced as an equivalent, not of the court masque, but of the genre of bird's-eye view engravings, etchings and drawings that had been popularised in 17th-century England primarily by Dutch artists steeped in the Flemish tradition of landscape painting. Topographical views from great heights remained *en vogue* in Stuart England – particularly for the 'total' representation of estates in the artificial mode of comprehensive bird's-eye views that tend towards map-making (see Wilton 43—46). This artistic convention, I argue, provides a complex framing, in specular terms, for Hobbes's hindsight account of the English Civil War and Cromwell's Protectorate. The first move consists of a putative equation of time and space, by way of which the former is made to take on some qualities of the latter: high and low. The assessment of actual events in time – i.e., the production of history – is rendered as land survey, if not cartography, so that the period in question gets transformed and circumscribed as a specific territory, a mountain ridge or high plateau that all the same retains its temporal essence: "the highest of time". Hobbes's interlocutor is after the spatialisation of history in order to produce this latter as a visible object available to the gaze as 'prospect'. What is introduced as a vista of course has to shade off into narrative so that a text similar to a 'literary map' emerges – a genre that, as Bernhard Klein has pointed out, formed an integral effect of early modern English map-culture, given the "susceptibility of the cartographic image to its ideological appropriation by the logic of narrative" (Klein 134). Even if Hobbes does not simulate a map but a mountain-top panorama, his text clearly plays on this permeability that, according to Klein, pertains between the visual and the textual 'maps'.

If the specular object that Hobbes introduces, however, has already been established as 'the highest', how can it lend itself at the same time to a spectator as prospect? Clearly, the spectator has to be positioned, in relation to his object, at an even higher point in space. Spatialised history, however, holds no place above 'the highest of time'. Accordingly, the impossible vantage point from which 'the period that passed between the years of 1640 and 1660' can be surveyed has to be located on an altogether different spatio-temporal grid: on the 'Devil's Mountain', the site of one of the three temptations of Christ according to Matthew and Luke. Interestingly, the King James Bible's rendition of the episode in Luke 4_5 already prefigures the conflation of time and space that Hobbes employs in his construction of history as territory: "And the devil, taking him [Jesus] up into an high mountain, shewed unto him all the kingdoms of the world in a moment of time". In this virtually global perspective, the time-bound process of scanning the surface of the earth is replaced by the flash-like vision of the totality, so that the temporality inherent in the act of seeing is eliminated in this instantaneous disclosure. The arrested synchrony ('in a moment of time') that enables and underwrites this comprehensive vision of 'all the kingdoms of the world' transforms empirical space into a tableau, if not a map, that renders the gaze itself acquisitive; hence Satan's imperial offer to Jesus is based on the contiguity of survey and possession: "All this power I will give thee, and the glory of them". In *Orientalism*, Edward Said describes imperialist vision as an integral part of conquest, a "static system of 'synchronic essentialism' [...] because it presumes that the whole Orient can be seen panoptically" (Said 240). From the Devil's Mountain according to Luke, not only the 'Orient' but the whole world is subjected to such synchronic essentialism; accordingly, it is Jesus's task to reject the panorama offered to him by the devil, and to identify the view from the 'high mountain' as a temptation to be resisted.

With what implications, and to what effects, then, does Hobbes locate his speaker – and by extension, his reader – on the Devil's Mountain as the one place from which the exceedingly 'high time' of the Civil War becomes available as prospect? Arguably, the conjuncture of secular and biblical history is designed to ensure that whatever offers itself to the gaze has to be read in terms of temptation. The frame therefore overcodes the entire subsequent narrative as a dangerous 'vista' that calls out for its own containment. Sure enough, the biblical 'original' asserts precisely such containment as Luke and Matthew, while differing in some respects, coincide with regards to Jesus's response to the devil's offer: "Then Jesus said unto him, Get thee hence Satan: for it is written, Thou shalt worship the Lord thy God, and him only shalt thou serve" (Matthew 4_{10}); "And Jesus answered and said unto him, Get thee behind me, Satan: For it is written, Thou shalt worship the Lord thy God, and him only shalt thou serve" (Luke 4_8).

Hobbes does not even have to unfold this narrative as he rightly expects his contemporary target audience to know it by heart[7]. Hence, the biblical account of the temptation of Christ would serve to prefigure and authorise the main theme of *Behemoth*, namely the distinction between legitimate (monarchical) and illegitimate (democratic) sovereignty: As Jesus refuses Satan's offer to be made ruler over 'all the kingdoms', and instead reasserts his subservience to the authority of God, likewise the logic of the implicit analogy in *Behemoth* would instil the reader's insight into the good citizen's "obligation towards the holder of power" – that crucial and immovable obligation that Hobbes had extensively laid down in both *On the Citizen* and *Leviathan*: "no number of citizens can

[7] This is in itself, of course, a point of complaint for Hobbes who emphasises time and again in *Behemoth* how devastating the effects of popular access to scripture are.

rightly strip the ruler of his power unless he gives his own consent as well" (Hobbes 1998: 90). Only the superior and ultimate authority of god might legitimise the removal of the sovereign; but that divine authority, as Hobbes points out time and again in his works, can only be claimed by fraud as in fact "there is no covenant with God, but by mediation of some body that representeth Gods person; which none does but Gods Lieutenant, who has the sovereignty under God" (Hobbes 1996: 122).

Hence, any claim to a legitimate alteration of the machinery of the State is by definition foreclosed: The divine law as laid down in scripture converges with Hobbes's version of the law of nature on the point of the supreme duty of individual obedience: "The whole of *the law of nature* is *divine,* and *conversely,* the whole of *the law of Christ* (fully laid out in chapters 5—7 of Matthew) is nature's teaching" (Hobbes 1998: 65). In order to buttress this synergy of immovably settled legitimacy, Hobbes has to disclaim all positions that purport to be derived from direct, 'lateral' communication with god. Neither "can any minister now say, that he hath immediately from God's own mouth received a command to disobey the King" (*B* 53); nor have 'fanatics' any legitimacy when they "pretend to speak with God and know his will by any other way than reading the Scriptures, which we also do" (*B* 90). The claim to political agency derived from individual and immediate "Divine inspiration" can only occur to "the great trouble of the Church and nation" (*B* 136). The decentralisation (or democratisation) of the access to the divine word sets in, for Hobbes, precisely with the vernacularisation of the scripture: "For after the Bible was translated into English, every man, nay, every boy and wench, that could read English, thought they spoke with God Almighty, and understood what he said" (*B* 21). Such dissemination and dispersal – such perspectivisation – of the erstwhile centralised word of the scripture effects the emergence of a culture of debate, of "the interpretation of the Scripture" (*B* 57), that Hobbes abhors as the hotbed of 'sedition'. The 'rule of opinion', the newly emergent public sphere of early print capitalist England[8] must, from Hobbes's Devil's Mountain, appear as a monstrous machine designed to manufacture dissent. In other words, hegemony takes centre stage in *Behemoth*, inasmuch as democratisation (of access to god, of bible exegesis, and concepts of State) affects the Hobbesian edifice at its core: The sovereign is in the last resort nothing but the product of the subject, and has

---

[8] Habermas locates the earliest emergence of a politically functional public sphere in late-17th-century England (Habermas 69-78). More recent work has utilised Habermas's model of the structural transformation of the public sphere in order to demonstrate how mid-17th-century is already marked by a vibrant and political culture of public debate (see Achinstein; Loewenstein; Zaret).

to rely on the latter's 'love of obedience'[9]. Hobbes's problem now is that the "state can constrain obedience, but convince no error, nor alter the minds of them that believe they have better reason" (*B* 62). In its struggle over hegemony, the state therefore has to contain the 'seditious' impulse at its source: It has to systematically foreclose the general access to the perspectival vantage point of scripture and interpretation. At this point, Hobbes resorts again to the imagery of the mountain, which he posits as an exclusive space accessible only to the authorised few. In this regard, even the pope – epithetically branded as 'tyrant' all through the narrative – can figure as a role model to be emulated by legitimate sovereign rulers:

> the Pope did concerning the Scripture the same that Moses did concerning Mount Sinai. Moses suffered no man to go up to it to hear God speak or gaze upon him, but such as he himself took with him; and the Pope suffered none to speak with God in the Scriptures, that had not some part of the Pope's spirit in him, for which he might be trusted (B 21).

From here, a profound instability begins to emerge in Hobbes's text; for if it is true that the account of the Civil War (which is after all what *Behemoth* is all about) should be read as the prospect "from the Devil's Mountain", then the entire narrative, as textualised perspectival vision, would have to be rejected as a satanic temptation. However, it is only in the form of a mountain-top panorama that this text can constitute itself: as that, then, which it prohibits. Hobbes's text, in other words, is political pornography, providing an occasion for showing that which is not supposed to be seen.

Blindness as sight: With John Milton on the top of speculation

As if to underscore its diametrical opposition to *Behemoth*'s authoritarianism right from the start, *Paradise Lost* opens with an invocation of precisely those figures that Hobbes heavily interdicts: divine inspiration and, more tangibly, the bird's-eye view. The conventionally addressed "Heav'nly Muse" (Milton 1993: 8)[10] gets immediately transposed from the domain of classical antiquity to that of judaeo-christian divinity; Milton's source of inspiration resides "on the secret top | Of Oreb, or of Sinai" (1:6-7), or else "Sion Hill" (1:10): elevated seats of privileged and unimpeded vision where "heav'n hides nothing from thy view" (1:27). Like *Behemoth*, then, *Paradise Lost* is framed as an enunciation from a bird's-eye perspective; Milton, however, has this perspective massively

[9] See, e.g., *B* 144: "It is not the right of the sovereign, though granted him by every man's express consent, that can enable him to do his office; it is the obedience of the subject, which must do that".

[10] Subsequently quoted in my text as book and verse numbers.

sanctioned by a divine authority that not only grants admission to precisely those altitudes that Hobbes has foreclosed, but that enables a humanistic agency that Hobbes goes out of his way to rule out: In Milton, seeing from a great height figures, not as a temptation but as the ascent "to the highth of this great argument", so that the speaker "may assert Eternal Providence | And justify the ways of God to man" (1:24-26). Milton's poem celebrates an unbridled vision that does not get revoked or contained by interdiction. It indeed exemplifies, as Marshall MacLuhan argues, one of the earliest literary simulations of three-dimensional, perspectivised space (MacLuhan 28). But – analogous to the abstract rationality of Renaissance perspective in general, and the conventions of the contemporary bird's-eye view in particular – this is, and has to be, a decidedly non-empirical vision: Like the author, the speaker of the poem is blind, as an autobiographical sequence at the opening of Book 3 indicates. Here, the speaker complains in an extended address to the "holy Light": that "thou | Revisit'st not these eyes that roll in vain | To find thy piercing ray, and find no dawn; | So thick a drop serene hath quenched their orbs, | Or dim suffusion veiled" (3:22-26). In other texts, Milton explicitly insists on an agency independent of physiological vision: most compactly in the 16th sonnet ("On his Blindness"), most radically in *Samson Agonistes*, most explicitly in the *Second Defence of the People of England*, where he polemically asserts, against his royalist 'defilers', to "be at once blind, and of the most piercing sight"; "not grown torpid by indolence since my eyes deserted me, but [...] still active, still ready to advance among the foremost to the most arduous struggles for liberty" (Milton 1999: 342; 343). Blindness, in *Paradise Lost*, appears to relegate vision to the prophetic mode that enables "to see and tell | Of things invisible to mortal sight" (3:54-55). Yet what this blind seer sees is primarily how others see the world in perspective.

Restricting the argument to the spectacular bird's-eye vision towards the ending of the poem, it becomes apparent how non-physiological panopticism indicates agentive subjectivity. In other words, the modern subject as which Adam is constituted, emerges first as an agent of sight, with the proviso that 'sight' does not necessitate the specular. Again, a mountain-top panorama is evoked as Adam, after the Fall and immediately before the expulsion from Eden, is led by the archangel Michael up an exceedingly high mountain "from whose top | The hemisphere of earth in clearest ken | Stretched out to amplest reach of prospect lay" (11:379-80). What ensues is the lengthy rendition of an unimpeded grand vista that breathtakingly encompasses geographical space from China via India and Arabia to Congo and Angola, and to Peru and Guyana – a celebration of seeing as such that wavers ambiguously between imperial fantasy and a well-nigh transmodern anti-Eurocentrism. The speaker makes it explicitly clear that, despite all analogies, this is neither the 'Devil's Mountain' nor a temptation:

"Not higher that hill nor wider looking round | Whereon *for different cause* the Tempter set | Our second Adam in the wilderness, | To show him all earth's kingdoms and their glory" (11:381-84; my emphasis). Yet the consistency of the assertion that we are not witness to imperialist temptation is open to debate. For once, the mountain-top vista *does* result in an acquisitive vision not quite innocent of early colonialist overtones; Milton's geography can for good reason be read as "a cataloguing and classification of space under the commanding eye of English standards – a process that privileges England whilst 'representing' a foreign geography" (McLeod 55). Read in this manner, Michael's revelation to Adam would – against the assertion that it happens *for a different cause* than imperial appropriation of the visual object – in fact coincide with Satan's presentation of 'all the kingdoms' to Jesus. Johannes Fabian has demonstrated, in a different context, how "synopticism – the urge to visualise a great multitude of pieces of information as orderly arrangements, systems, and *tableaux* – [forms] a constant temptation" (Fabian 118). What, however, if the grand panorama in Milton's narrative is never offered to Adam but only referred to as part of the allusion to the biblical temptation episode? Then the enumeration of exotic 'seats of power' would add up to one long digression illustrating the episode of Jesus and Satan that is juxtaposed to the main narrative of Adam and Michael. Most critics have overlooked the conspicuous indeterminacy of the passage, as the clear, assertive diegesis gives way to (deliberately?) vague speculation. For immediately subsequent to the reference to the temptation of Christ, the pronouns and deictic markers begin to blur:

> Not higher that hill nor wider looking round
> Whereon for different cause the Tempter set
> Our second Adam in the wilderness,
> To show him all earth's kingdoms and their glory.
> *His* eye might *there* command wherever stood
> Cities of old or modern fame, the seat
> Of mightiest empire (11:381-387; my emphases).

This seems to me an instance of that "fluidity of Milton's syntax" (Ricks 81) which Christopher Ricks has demonstrated more than forty years ago: One of the most striking effects of the 'liquid texture' of *Paradise Lost*, according to Ricks, consists in "leaving it possible for a word or a clause to look backward or forward" (Ricks 137) at the same time. Such fluidity, of course, invites to be read in terms of form and style as sedimented content, so that the grammatical undecidability of Milton's sentences would correlate with ideological incertitude. *Paradise Lost* as an ultimately aporetic text would on this reading emerge as deeply informed by "indeterminacy and difference of opinions" (Rumrich 37). More recently, Peter C. Herman has assigned to Milton a "poetics of *incertitude* because it is out of the turmoil of not knowing what to affirm in the wake of the Revolution's failure that Milton creates his finest poetry" (Herman 183). At a modest but not insignificant scale, it is ultimately impossible to decide who is the 'he', where the 'there' of verse 385 in the sequence quoted above – Adam under Michael's tutorship on the highest hill of Eden, or Jesus tempted on the 'Devil's Mountain'? To whom, and from which point, is the ensuing panorama of enumerated imperial 'seats' from 'Cambalu' to 'El Dorado' revealed? The long list of exotic centres of power, past and present, cannot help to clarify this, especially if one considers the resolution of this global panorama: "but to nobler sights | Michael from Adam's eyes the film removed | Which that false fruit that promised clearer sight | Had bred" (11:411—414). At first sight, this angelic intervention seems to occur *after* Adam's perusal of the Renaissance world in the manner of a map; even then, it is obviously underscored that this is a vision influenced by the 'false fruit' of the Fall. Abiding by Milton's narrative, however, it is much more plausible to assume that Adam – other than Jesus – is in fact never granted (or tempted with) this geographic vision, and that the 'removal of the film' is necessary, not for the exorcism of the imperial desire inherent in the geopolitical gaze (which we now read as in fact withheld from Adam), but in order to prepare Adam for the 'nobler sights' lying in waiting for him. Instead of impossibly compressed space, then, Adam is offered compressed time. For it is indeed the entirety of human history from Cain to the second coming that Michael reveals to Adam – first as drug-induced spectacular pageant, then, after Adam's "mortal sight" is exhausted, as lecture:

> Much thou hast yet to see, but I perceive
> Thy mortal sight to fail; objects divine
> Must needs impair and weary human sense:
> Henceforth what is to come I will relate
> Thou therefore give due audience, and attend. (12:8-12)

The exhaustion of mortal sight – blindness – does not cancel the great vision: After all, this is the simulation of three-dimensional sight with the strictly non-visual means of writing[11]. Within the diegesis of *Paradise Lost*, moreover, Michael's imperative announces a switch of genre (from pageant to verbal narrative) which ensures that Adam's failure of physiological sight does not impair his 'seeing' even though reception shifts from vision to 'due audience'. In fact, Michael's long narrative repeatedly takes on the features of visual demonstration in perspectivised space: "each place behold | In prospect as I point them" (12:142-43). Accordingly, at the stage when Michael's narrative has reached the return of the Israelites to Canaan, the 'blind' Adam can interpolate

[11] For a principal discussion of language's incapacity to (literally) 'depict', see Seymour Chatman, "What Novels Can Do that Films Can't (And Vice Versa)".

an ejaculation of non-specular scopophilia: "Enlight'ner of my darkness, gracious things | Thou hast revealed [...] | [...] now first I find | My eyes true op'ning" (12:271-274). As showing gives way to telling, the mountain is transformed from a specular vantage point to the "top | Of speculation" (12:588-89): a location that the reader of the poem is invited, if not impelled, to co-occupy along with Adam in order to imbibe the instructive angelic verses that produce a merely verbal vision. Where Hobbes's panoramic account of the Civil War has to be incriminated as pornography, Milton's much wider vistas, encompassing the entirety of the "Measured [...] transient world" (12:553), reveal the totality of legitimate knowledge, which is circumscribed and delimited only by the intangible vanishing point of perspective itself: "beyond is all abyss, | Eternity, whose end no eye can reach" (12:555-56).

On the one hand, Michael's instruction thus contains the whole of the knowable world defined by the reach of perspective as continuous, homogeneous, rational time/space. On the other, this apparently strictly immanentist prospect is spatially framed by the anti-perspectival 'abyss' of eternity, just like the history that Michael unfolds is informed by a transhistorical providential subtext. The knowable world is therefore intimately suffused with eschatology, and the angelic presentation adds up to a lesson in good Christian conduct according to which "to obey is best, | And love with fear the only God, to walk | As in his presence, ever to observe | His providence, and on him sole depend" (12:561-64). The supreme imperative of obedience to god, however, entails – and this is where Milton remains stubbornly 'republican' even after the failure of the revolution – an unwavering compliance with the divine decree of "fair equality" among humans (12:26). With the insistence on a strictly divine prerogative to the title of lord, *Paradise Lost* dramatises a revaluation of sedition itself. The biblical Nimrod, as the first man to pretend to the title of king, figures as the archetypal 'rebel' against the original political order decreed by god, who "man over man | [...] made not lord; such title to himself | Reserving, human left from human free" (12:69-71). The institution of monarchy, therefore, is based on "Authority usurped, from God not given" (12:66), and assigns a mere human the status of god himself. Hobbes, famously, had designed the artificial person/actor of the absolute sovereign as "that *Mortall God*, to which wee owe under the *Immortall God* our peace and defence" (Hobbes 1996: 120). For Milton, this is not only an offence to "Rational liberty" (12:82) but the blasphemy inherent in an ideology that substitutes man for god, and justifies the tyrant who is "despite of Heav'n | [...] claiming second sovranty; | And from rebellion shall derive his name, | Though of rebellion others he accuse" (12:35-37).

This Whiggish version of a rationally/theologically republican Milton, of course, is highly selective. Milton's republicanism is, like that of so many of his humanistic contemporaries, derived from a specific reading of classical political theory in which the authoritative figure of the instructor replaces the royalist figure of the king. Milton's political prose is replete with fantasies of elevating a nation, by way of instruction, to the standard of 'civility' required for a free state at home, and imperial claims abroad. His own post-restoration account of the Long Parliament – the 'Digression' that supplements his *History of Britain* – explicates the thorough disillusionment of this political pedagogy: The course of the revolution leaves the British, even when they gain military victories over internal and external opponents, "under the same grievances that men suffer conquered" (Milton 1991: 11). Why? Because the nation proves incapable of performing the general civilisational progress that Milton had envisaged, and even projected as actually underway, in such euphoric radical democratic tracts as *Areopagitica*. Instead of the optimistic invocation of a "puissant nation rousing herself like a strong man after sleep, and shaking her invincible locks" (Milton 1999: 43), the 1681 'Digression' ends on a note of historical reversal. On the debris of the Protectorate, Milton asserts the structural affinity of the English revolutionaries with their incompetent historical forebears, the uncivilised, 'barbarous' ancient Britons:

> From the confluence of all their [the Parliamentarians'] Errors, Mischiefs, and Misdemeanous, what in the eyes of Man could be expected, but what befel those Antient Inhabitants whom they so much resembled, Confusion in the end? (Milton 1991: 11).

The elitism of Milton's republicanism comes aggressively to the fore at the moment of disappointment with the lack of civility in his contemporary countrymen. In *Paradise Lost*, this historical experience of defeat is anchored in an existential, post-lapsarian human condition. When Michael states that, "Since thy original lapse, true liberty | Is lost" (12:83-84), this framing of the fallen state of humanity prepares the ground for the apparently un-Miltonic political assertion that "tyranny must be" (12:95). Surprisingly, Milton's Michael argues in a decidedly Hobbesian fashion when he explains the necessity of tyranny with the 'natural state' of man after the Fall:

> Reason in man obscured, or not obeyed
> Immediately inordinate desires
> And upstart passions catch the government
> From reason [...]
> [...] Therefore since he permits
> Within himself unworthy powers to reign
> Over free reason, God in judgment just
> Subjects him from without to violent lords (12:86-93).

This, to be sure, is not pedagogy but 'judgment just': The establishment of tyranny, though originally a rebellion against the 'natural' equality of humans *and* the divine prerogative to the title of lord, is strictly homologous to man's inner subjection to 'upstart passions' – a false 'government' of which political tyranny forms the external equivalent and logical extension[12]. As in Hobbes, then, it seems like it were the subject that produces the sovereign as tyrant. For Milton, however, this remains an exceptional state of collective degradation (which he is ready to apply to restoration England): "Yet sometimes nations will decline so low | From virtue which is reason, that no wrong, | But justice [...] | Deprives them of their outward liberty | Their inward lost" (12:97-101). Inwardness, one of the great inventions of the English Renaissance (see Maus), finally emerges in Milton's text as the ultimate site of agency. Interiorisation of the gaze organises the shift from pageant to lecture, from specular vantage point to top of speculation: What is 'taken in' by Adam's non-seeing eye results in nothing less than the insight into the necessity of obedience as precondition of freedom. The instruction through Michael is accordingly resumed with the substitution of a "paradise within [...], happier far" (12:587) for the exteriority of Eden itself. Accordingly, when Adam and Eve finally evacuate the garden, they enter the "world [that] was all before them" (12:646) as subjects fully equipped with an interiority tuned to the Christian cosmos. *Paradise Lost* closes with the beginning of post-lapsarian human history in a condition that qualifies men and women for, and entitles them to, the role of free – that is, obedient – political agents.

Works Cited

Achinstein, Sharon (1992). "The Politcs of Babel in the English Revolution". *Pamphlet Wars: Prose in the English Revolution.* Ed. James Holstun. London: Frank Cass, 14-43.

Benjamin, Walter (1968 [1939]). "Theses on the Philosophy of History". *Illuminations: Essays and Reflections.* Ed. & intr. Hannah Arendt. Tr. Harry Zohn. New York: Schocken, 253-264.

Berger , John (1997 [1972]). *Ways of Seeing.* Harmondsworth: Penguin.

Bunyan, John (1984 [1678]). *The Pilgrim's Progress.* Ed. & intr. Roger Sharrock. Harmondsworth: Penguin.

Cassirer, Ernst (1965 [1925]). *Philosophy of Symbolic Forms. Vol 2: Mythical Thinking.* New Haven: Yale UP.

Chatman, Seymour (1992 [1980]). "What Novels Can Do That Films Can't (And Vice Versa)". *Film Theory and Criticism.* Fourth Edition. Ed. Gerald Mast, Marshall Cohen & Leo Braudy. Oxford: OUP 403-419.

---

[12] Herman Rapaport goes as far as charging Milton of "complicity with the most repellent aspects of fascism"; with Milton, Rapaport argues, "the modern totalitarian state has made its hideous entry. The Third Reich speaks *avant la lettre*" (Rapaport: 172; 177).

Chatterjee, Partha (2003). "Anderson's Utopia". *Grounds of Comparison: Around the Work of Benedict Anderson*. Ed. Jonathan Culler & Pheng Chaeh. New York & London: Routledge, 161-170.

Fabian, Johannes (1983). *Time and the Other: How Anthropology Makes Its Object*. New York: Columbia UP.

Habermas, Jürgen (1968 [1962]). *Strukturwandel der Öffentlichkeit: Untersuchungen zu einer Kategorie der bürgerlichen Gesellschaft*. Neuwied: Luchterhand.

Herman, Peter C. (2003). "*Paradise Lost*, the Miltonic 'Or', and the Poetics of Incertitude". *SEL* 43.1, 181-211.

Hill, Christopher (1972). *The World Turned Upside Down: Radical Ideas in the English Revolution*. London: Temple Smith.

Hobbes, Thomas (1998 [1641]). *On the Citizen (De cive)*. Ed. & tr. Richard Tuck. Cambridge: CUP.

Hobbes, Thomas (1996 [1651]). *Leviathan*. Ed. Richard Tuck. Cambridge: CUP.

Hobbes, Thomas (1990 [1682]). *Behemoth; or, The Long Parliament*. Ed. Ferdinand Tönnies. Intr. Stephen Holmes. Chicago & London: U of Chicago P.

Holly, Michael Ann (1992). "Writing Leonardo Backwards". *New Literary History* 23, 173-211.

Holstun, James (2000). *Ehud's Dagger: Class Struggle in the English Revolution*. London & New York: Verso.

Jameson, Fredric (1981). *The Political Unconscious: Narrative as a Socially Symbolic Act*. London: Methuen.

Kaufmann, Thomas DaCosta (1993). *The Mastery of Nature: Aspects of Art, Science, and Humanism in the Renaissance*. Princeton: Princeton UP.

Klein, Bernhard (2001). *Maps and the Writing of Space in Early Modern England*. Houndmills: Palgrave.

Loewenstein, David (2001). *Representing Revolution in Milton and his Contemporaries: Religion, Politics, and Polemics in Radical Puritanism*. Cambridge: CUP.

Maus, Katharine Eisman (1995). *Inwardness and Theater in the English Renaissance*. Chicago: Chicago UP.

McLeod, Bruce (1999). "The 'Lordly Eye': Milton and the Strategic Geography of Empire". *Milton and the Imperial Vision*. Ed. Balachandra Rajan & Elizabeth Sauer. Pittsburgh/Pa.: Duquesne UP, 48-66.

McLuhan Marshall (2002 [1962]). *The Gutenberg Galaxy: The Making of Typographic Man*. Toronto: U of Toronto P.

Milton, John (1999 [1644]). *Areopagitica. Areopagitica and Other Political Writjngs of John Milton*. Intr. John Alvis. Indianapolis: Liberty Fund, 1-51.

Milton, John (1999 [1654]). "The Second Defence of the People of England". *Areopagitica, and Other Political Writings of John Milton*. Ed. John Alvis. Indianapolis: Liberty Fund, 314-413.

Milton, John (1993 [1674]). *Paradise Lost*. Ed. Scott Elledge. New York: Norton.

Milton, John (1991 [1681]). "Mr John Milton's Character of the Long Parliament in 1641". *The History of Britain*. Appendix 1. Intr. Graham Parry. Stamford: Watkins.

Mitchell, W. J. Thomas (1986). *Iconology: Image, Text, Ideology*. Chicago & London: Chicago UP.

Morgan, Victor (1983). "The Literary Image of Globes and Maps in Early Modern England". *English Map-Making 1500—1650: Historical Essays*. Ed. Sarah Tyacke. London: British Library, 46-55.

Panofsky, Erwin (1980 [1964]). "Die Perspektive als 'symbolische Form'". *Aufsätze zu Grundfragen der Kunstwissenschaft*. Ed. Hariolf Oberer & Egon Verheyen. Berlin: Spiess, 99-167.

Rapaport, Herman (1983). *Milton and the Postmodern*. Lincoln & London: U of Nebraska P.

Ricks, Christopher (1963). *Milton's Grand Style*. Oxford: Clarendon Press.

Rumrich, John P. (1996). *Milton Unbound: Controversy and Reinterpretation*. Cambridge: CUP.

Said, Edward W. (1994 [1978]). *Orientalism*. New York: Vintage.

Schmeiser, Leonhard (2002). *Die Erfindung der Zentralperspektive und die Entstehung der neuzeitlichen Wissenschaft*. München: Fink.

Sinfield, Alan (1992). *Faultlines: Cultural Materialism and the Politics of Dissident Reading*. Oxford: Clarendon Press.

Tyacke, Sarah (2001). *Before Empire: The English World Picture in the Sixteenth and Early Seventeenth Centuries*. London: Hakluyt Society.

Wilton, Andrew (2001). *Five Centuries of British Painting: From Holbein to Hodgkin*. London: Thames & Hudson.

Zaret, David (2000). *Origins of Democratic Culture: Printing, Petitions, and the Public Sphere in Early-Modern England*. Princeton: Princeton UP.

Jatin Waglé

Resistant Forms and Critical Subjectivities

Reading Adorno in Our Times and Other Languages

Darum ist das innerste Formgesetz des Essays die Ketzerei. (Adorno 1991b: 33)[1]
Aber Philosophie ist der Einspruch gegen das Äquivalenzprinzip... (Adorno 1991a: 100)[2]

1 Introduction

It is by now a truism that the question of discursive resistance is tied up with that of agency in utterances, scholarly or aesthetic. However, there appears to be no unanimity on what may constitute such textual/political opposition. The texts that defy contemporary hermeneutic frames may be characterised as resistant. But, this is a vexed issue, because a wide array of approaches could be classified as resistant by many for several reasons. However, what appears to be common to them is their somewhat predictable labelling as 'difficult'. A diverse range of methods and strategies of stating, enunciating and writing are experienced as being demanding or challenging by readers. From amongst those who are famed for their difficulty, and whose *oeuvre* challenges habitual readings, one could cite disparate figures – those who structure theory and those who compose fiction – such as Theodor Wiesengrund Adorno, James Joyce, Arno Schmidt, Jacques Lacan and Jacques Derrida. This sampling mixes the discursive with the aesthetic deliberately, the suggestion being, when it comes to the practice of discursive resistance, there is much that is common to the writings, whichever the mode. Although the deeper hermeneutic *Problematik* may be varied and illusive, one may identify the more accessible, common aspects: firstly, an implicit or explicit theory of language and of reading, a preoccupation with what may constitute resistance in writing and presentation, in other words, a theory of writing, and lastly, sustained, self-conscious experimentation with forms of expression.

[1] Hence the essay's innermost formal law is heresy (Adorno 1991c: 23).
[2] But philosophy is a protest against the principle of equivalence... (Adorno 1993: 108-9).

1.1 The Question of Form

Thus, there seems to be some sort of agreement across approaches that the question of resistance of and in discourses cannot be separated from their treatment of form. However, it may be observed that the experimentation with newer, more self-aware, forms of writing is a process that is subject to history. In other words, it may be argued that the oppositional strategies of a particular historical phase may be absorbed in other times, and those developed in particular contexts may be rendered complicit in changed cultural and linguistic environs. Furthermore, linguistic/cultural strategies of subversion might petrify in the face of an untoward moment, freeze over, and turn into innocuous fashions, or worse still, rigid fetishes. Therefore, a nuanced theory of forms of (re)presentations is needed to comprehend the complex politics of writing and reading resistance. Such a possibility, I believe, is available in the prolific and multifarious *oeuvre* of Theodor W. Adorno. The remarkable and enduring aspect of Adorno's writings, which sought to resist the discourses of complicity with either Fascism or modern Capitalism or both, is that they are deeply aware that a harmonious unity of language and meaning, form and thought is impossible. Nevertheless, they are also conscious that an analytical separation of form and content is no guarantee of resistance, and that seemingly progressive content is routinely subverted by naturalised forms.

1.1.1. Forms of Presentation in Adorno and the Anglophone Adorno-Discourse

In this short essay, I seek to address the praxis of resistant forms in Adorno's complex *oeuvre*, a body of work that was meant to challenge glib, ahistorical readings and defy unproblematic cooption. In the process, I wish to decipher the broad outlines of Adorno's theory of discursive resistance. While addressing Adorno's elaboration of resistant presentational forms [*Darstellungsformen*], I wish to briefly contextualise his work both historically as well as discursively. Secondly, I seek to extend my observations to the English language reception of Adorno's multifaceted *oeuvre*. The sizeable body of Anglophone Adorno-criticism does not address the question of presentational forms frontally. However, it is this very question regarding the form and presentation of writing that appears to constitute a vital moment of crisis in this discourse (Nicholsen 1997: 3). Quite a few commentators cite the question of Adorno's 'difficult', 'untranslatable' and 'impervious' forms as a central reason for his allegedly inadequate and frequently flawed translation and Anglo-American reception (Weber 1997: 11; Jameson 1990: ix; Nicholsen 1997: 2). In this context, I seek to read the aporia of the 'translation' of Adorno's *Darstellungsformen* into Anglophone second-order criticism as a productive forcefield, where questions of discursive resistance may be reread in terms of alterity and intercultural encounter.

2 Adorno's Theory of Writing

2.1 Contextualising Adorno's Resistant Writings

In Adorno's career as thinker and author, concerns about the relationship between content and form and what may constitute the language of critical thought are articulated very early. Moreover, he continued to reflect on and practise complex and oppositional forms that could represent his particular vision of Critical Theory (Hohendahl 1995: 218-25). Thus, Adorno's theory of resistant *Darstellungsformen* emerges as an anti-systematic, self-reflexive elaboration through his writings in diverse fields, such as philosophy, cultural criticism and aesthetics. Whereas his philosophising is remarkable for its relentless insistence on a dialectic that shuns the Idealist reconciliation (Adorno 2001), his aesthetics constitutes an attempt to immerse itself in the art-object(s) (Hullot-Kentor 1997: xii). In his *oeuvre*, conventional forms of writing are eschewed for configurational or constellational ones that seek to voice a content that is otherwise marginalised and repressed. Where experimentation with syntactical structures is concerned, across the range of his writings, one notices an inexorable emphasis on paratactical constructions which seek to preserve an equitable balance not just of ideas but also of their linguistic performances (Weber 1997). Peter Uwe Hohendahl (1995: 220) contextualises Adorno's project as "a relentless self-conscious critique of ossification as it is reflected in the language of positivism or ontology", or even of the more orthodox strands within the Marxist tradition.

2.2 Mapping Adorno's Theory of Discursive Resistance

Three significant aspects regarding Adorno's rigorous relationship with the question of discursive forms of resistance may be noted. Firstly, he worked with a somewhat idiosyncratic theory of language, which is comparable in its emphases to his theories of art and culture. Although he recognized the significance of the fundamental and everyday aspect of communication in language, he was extremely critical of what he saw as its growing complicity with the overarching systems of domination. He counterposed a nuanced and complexly framed theory of poetic or literary language that could resist the lure of commerce. In the process, he also attempted to resuscitate some of the more archaic potentialities of the German language, presumably to construct non-privileging linguistic structures. As Samuel Weber (1997: 14) puts it in his introduction to *Prisms*, "Adorno's language, constantly struggling with the communicative aspect of German, wrests its meaning from the latent potential which still inheres in German..." Secondly, he honed forms which challenged habituated, naturalized readings, and thereby, interrogated the reader's unproblematic relationship with words, texts and readings. To this effect, he worked with genres conventionally understood as being unscholarly or non-academic. His experiments, for instance, with the essay and the aphorism are known for their poetic eloquence as also their cryptic difficulty. More significantly, Adorno's self-reflexive employment of these formal strategies and his reflections on the subject articulate a non-systematic theory of resistant writing. Thirdly, his quest was to shape forms that spoke for and through the particular, and approximated to the non-identical, the aconceptual and the aesthetic. In the process, he developed an idiom, *Adorno-Deutsch*, famed for its intractable difficulty. As Shierry Weber Nicholsen (Adorno 1991c: x-xi) suggests, Adorno's writings verge on an artifice – a constructed art-language or *Kunstsprache* that frustrates, without explanation, expectations, and established patterns of thought and expression.

2.3 Adorno's Theory of Language

Although Adorno did not state a systematic theory of language[3], his whole *oeuvre* could be seen as an articulation of an implicit philosophy of language and presentational forms, which is inseparable from his major substantive utterances. According to Adorno (1992: 194), language is an intertwinement of two distinct yet inseparable dimensions – the 'intentional' language of communication and commerce with its affinity to the logic of domination, and an authentic, poetic, or expressive dimension which is threatened by the former and yet seeks to elude it. Thus, he saw his own writing as a political struggle against the gradual and possibly inevitable disintegration of language. According to one of Adorno's earliest translators, Samuel Weber (1997: 14), "If Adorno appears to do violence to ordinary German, it is as shock therapy which legitimizes itself in exposing the violence that language has already inflicted upon itself". Yet, it needs to be emphasized that Adorno did not succumb to any illusions regarding the so called natural purity that language might be restored to. Moreover, he was deeply aware of the complex implication of the 'mimetic' or aesthetic facet of language with its instrumental character. Not unlike several of his other dialectical formulations, Adorno's intricate position on language charts a seeming contradiction. According to him, as a mediated, historical

---

[3] For insightful attempts to uncover an unstated theory of language through close readings of Adorno's writings, see the chapters "The Discourse of Philosophy and the Problem of Language" in Hohendahl (1995: 217-42), and "Language: Its Murmurings, Its Darkness, and Its Silver Rib", in Nicholsen (1997: 59-102).

experience, language would always maintain its dual, divided self. However much literary language became hermetic in its efforts to shake off the language of exchange, it would never be truly free from it (Nicholsen 1997: 68-9).

2.3.1 Mimetic Language and Its Tenebrous Illumination

In his essays on the poetry of Eichendorff (1991c: 55-79) and Rudolf Borchardt (1992: 193-210), Adorno drafts a complex formulation regarding the awakening of the aesthetic potential in language. He suggests that when the self willingly extinguishes itself in language, there may emerge a fleeting, 'homeless' language of 'murmuring' [*Rauschen*] which turns into 'second nature', and in which "the objectified nature that has been lost to the subject returns as animated nature" (Adorno 1991c: 69). This state of 'the darkness of language' is not only a movement of language away from subjectivity and communication "toward the vanishing Other, the nonidentical object" (Nicholsen 1997: 89), but also a renunciation of violence towards language. According to Adorno, this violence that language forces upon itself is an inevitable facet of the instrumentality and communicability inherent in the 'ordinary' life of language. On the other hand, in this state of authentic linguistic being, meaning understood as communicable equivalence gives way to mimetic import or substance [*Gehalt*], as the language of murmuring approaches music. This 'nonviolent language' – a product of self-alienation [*Selbstentäusserung*] – "succeeds in... washing words away from their circumscribed meanings and causing them to light up when they come in contact with one another" (Adorno 1991c: 70). For Adorno, this dialectical process – involving the movement of transcendence through negation and reconciliation – leads to the configurational form, combining the indistinctness of gaps and breaks in language with the "constitutive disassociation" of parataxis (Adorno 1992: 130).

2.4 Language of Anti-Systematic Philosophising

In his complex piece on the question of intelligibility in Hegel's allegedly obscure writings, "Skoteinos, or How to Read Hegel", Adorno (1993b: 89-148), while delineating a theory of reading, moves into a broader theory regarding the writing of philosophical prose. In other words, he seems to extend his theory of language to that of the language of philosophy. While commenting on the inadvertent, hidden processes through which Hegel's writing style takes on a music-like quality, he suggests that the inherent nature of philosophy should actually bring it closer to the literary, 'mimetic' rather than the intentional, communicative dimension of language: "Philosophy as a whole is allied with art in wanting to rescue, in the medium of the concept, the mimesis that the concept represses..." (Adorno 1993b: 123). Nevertheless, in the same essay Adorno launches a scathing critique of what he calls the "antilinguistic impulse" in Hegel. As he puts it, "This man who reflected on all reflection did not reflect on language" (Adorno 1993b: 122). Adorno's radically revisionist reading of Hegel goes on to suggest that Hegel's indifferent attitude towards language, his misplaced reliance on the concept of totality and his attempt to approximate his thought to a system, impart to his language the severity of a doctrine. Adorno counterposes the configurational or constellational form and its anti-systematic aspects as a mode of presentation that arrests the lapse of the dialectic into the doctrinaire:

> The specificity of philosophy as a configuration of moments is qualitatively different from a lack of ambiguity in every particular moment, even within the configuration, because the configuration itself is more, and other, than the quintessence of its moments. Constellation is not system. Everything does not become resolved, everything does not come out even; rather, one moment sheds light on the other… (Adorno 1993b: 109).

Furthermore, as a contrast to the Hegelian abstraction, Adorno puts forth the notion of concreteness – a term that recurs throughout his *oeuvre* – while discussing what should be the authentic agenda of contemporary philosophy or the immersion of aesthetics in its object. According to Weber (2002: 381-2; emphases retained), "…concreteness consists essentially in the *growing together* of thought through which knowledge is constituted. Such concretion involves the move beyond the conception of knowledge as passive *medium* to a conception of knowledge as an active *process of self-constituting*".

2.4.1 Essaying the Nonidentical

Adorno's construction of anti-systematic, philosophic discourse suggests three different yet converging movements, towards aesthetics or mimetic prose, towards open-ended constellational forms, and towards 'concrete' language. His insistence on these elements finds its most coherent articulation in his famed piece on "The Essay as Form" (1991c: 3-23). He reclaims this unscholarly form and defends it for its indirect and paratactical expression of the consciousness of nonidentity, its radical refusal to reduce its substance to a principle, and its fragmentariness. Above all, he champions this genre for its rebellion against the dogma that sees the individual, transitory, and temporal experience as being unworthy of philosophical enquiry. Adorno's theory of philosophical discourse should be seen, in this context, as an integral dimension of his rejection of the philosophy of the first principles. The poverty of systematic philosophising, he suggests, lies in its attempt to gloss over the mediation between the particular historical experience and its rendering of it as an abstract totality. Therefore, he is deeply sceptical of any attempt to grant primacy to the conceptual vis-à-vis the individual object of its abstraction. Adorno advocates the essay as the apposite vehicle for critical philosophical discourse, as it "challenges the notion that what has been produced historically is not a fit object of theory" (Adorno 1991c: 10). It could, therefore, be inferred that the essay for Adorno is a 'low' fragmentary form that eschews the grandstanding, totalizing abstraction of conventional high philosophising and yet retains the suppleness to articulate with complexity the 'ordinariness' of contemporary experience.

2.4.2 Theorising the Aconceptual

This trenchant questioning primarily – but not solely – of the language, styles and presentational forms employed by the philosophy of Idealism is echoed in the *Negative Dialectics* (Adorno 2001; 1973). A significant, but perhaps undervalued, aspect of the text is its struggle to philosophise that which is considered unworthy of its attention by conventional philosophy. Adorno's endeavour to theorise the philosophical remnants begins again with a critique of Hegel. He states that the Hegelian "lazy existence"[4] is invariably excluded from one's awareness and the scripting of philosophy. According to this formulation, conventional theorizing excludes inexorably, as the non-identical is subsumed under the identity of the concept. As a part of his critique of traditional philosophy's disregard for the non-identical, and while scripting an agenda for a philosophy of the future, Adorno goes on to write, "Its theme would be the qualities which it has degraded to the merely contingent, to *quantité négligeable*" (Adorno 2001; 1973: 8). However, his critique of traditional philosophy is in the main a critique of the conceptual vocabulary that lies at its foundation. In order to salvage the aconceptual residue of philosophy, Adorno wages an immanent critique of the conceptual terrain. With a near impossible dialectic turn, he seeks to construct a conceptual paradigm that may respond to what it has denied without subsuming it: "Whatever of the truth can be gleaned through concepts beyond their abstract circumference, can have no other staging-grounds than that which is suppressed, disparaged and thrown away by concepts. The utopia of cognition would be to open up the non-conceptual with concepts, without making it the same as them" (Adorno 2001; 1973: 9).

2.4.2.1 Salvaging the Hegelian Lazy Existence

In *Negative Dialectics*, Adorno outlines the possibilities of reclaiming the lazy existence and of formulating an immanent critique on its behalf – an intricate

[4] According to Adorno, this is Hegel's pejorative label for "the non-conceptual, the individual and the particular", which is dismissed in conventional philosophy as transient and inconsequential, and which should be the true interest of philosophy of our times (Adorno 2001; 1973b: 8).

process that entails the splitting up of the dominant concept and its identity with the subject, and where the subjugated nonidentical harnesses its apparent weakness to break the magical spell of identity. While critiquing the Hegelian dialectic and mapping out the path of a 'negative dialectic' which eschews easy reconciliation, Adorno claims,

> Its movement does not tend towards identity in the difference of every object from its concept; rather it suspects that which is identical. Its logic is one of disassembly [Zerfall]: of the well-equipped and concretized form of concepts, which the cognizing subject immediately faces at first. Their identity with the subject is the untruth. With this untruth, the subjective pre-formation of the phenomenon slides in front of what is non-identical, before the *individuum ineffabile*. ...In the process of demythologization positivity must be negated all the way to the instrumental reason, which secures demythologization. ...The subject must render compensation to the non-identical, for what it perpetrated on it (Adorno 2001; 1973:145; translation modified.).

2.5 Primacy of the Object

Adorno's avowal of resistant, fragmentary forms and his emphasis on a philosophical vocabulary that speaks on behalf of the particular and the transitional stem from his unwavering insistence on the 'primacy of the object' [*Vorrang des Objekts*]. With the dominating rationality corresponding to the extant 'totally administered society', Adorno sought a utopian, alternative form of aconceptual rationality. According to Robert Hullot-Kentor (2004: 182), Adorno's philosophy, if it could be conceived of as a whole, seeks the primacy of the object. His critique of German Idealism in general and Hegel's philosophic system in particular – in terms of the privileging of the concept and the identity of the subject – leads to the theory of the primacy of the object, which in turn forms the kernel of his 'nonidentitarian thinking' (Adorno 1973; 1993b; Weyand 2001: 23-6). Similarly, where his writings on art and literature are concerned, he seeks to immerse his critique in its object. Adorno's aesthetic theory attempts to be an immanent articulation of the art-object, the self-understanding of art, or, even the revelation of its truth content [*Wahrheitsgehalt*] (Hullot-Kentor 1997: xii, xiii).

3 Anglophone Adorno-Discourse and the Question of Forms

Adorno's English language reception appears to flounder on the question of the theory and praxis of resistant forms in his *oeuvre*. Nicholsen (1997: 3), for instance, claims that a critical problematic facet of Anglophone Adorno-criticism has been its underestimation of the "centrality of the aesthetic dimension" in his *oeuvre*, and by extension, its neglect of his insistence on the "presentational form [*Darstellungsform*] of his work". Therefore, the hubris of Anglophone Adorno-scholarship is its undervaluing of the project of the primacy of the object, which was pivotal in his *oeuvre*. Nicholsen (Ibid.) recommends that to locate Adorno's true actuality for the present one must delve deeper into the connection drawn in his work "between the aesthetic dimension and a nondiscursive form of truth". Thus, it is ironically the question of form in Adorno that would perhaps constitute a relevant substantive ground of future Adorno-research.

3.1 Framing the English-Language Adorno-Discourse

Although the second-order English language Adorno-criticism does not accord centrality to the question of presentational and linguistic forms in his writings, it is possible to draw out at least three broad strands of criticism that respond to the question in a more or less direct manner. Hohendahl (1995: 217) suggests that the Anglophone reception of Adorno's theory of language and writing may be viewed in the context of the post-Adornian linguistic turn in cultural theory, of both the Habermasian as well as the poststructuralist variety.

3.1.1 The Post-Habermasian Critique

The first approach follows the critical insights developed by Jürgen Habermas (1983b; 1987b) and then by Albrecht Wellmer (1991: 1-35). Habermas' post-Adornian Critical Theory perceives itself as the "rejection of the 'paradigm of consciousness' and its associated 'philosophy of the subject' in favour of the through-and-through intersubjectivist paradigm of 'communicative action'" (McCarthy 1987: x). In this context, the "rigid features" of Adorno's aesthetics are cited as a manifestation of what Wellmer identifies as a flawed, anachronistic notion of autonomous subjectivity in Adorno. Furthermore, Adorno's writings are seen as the "univocal" representations of a lonely subject unwilling or unable to converse with other voices. However, few Anglophone critics appear to subscribe to the Wellmerian solution of redrafting Adorno's theory of forms on the Habermasian intersubjective-linguistic coordinates. Moreover, for all its awareness of a post-Adornian, multivocal intersubjectivity, the post-Habermasian academic prose appears to be rather unselfconscious in its deployment of formal strategies and hardly, if ever, defiant of the naturalized reading positions.

3.1.2 The Postmodernist Rejection

In the postmodernist response to the question of discursive forms in Adorno, the Wellmerian critique regarding the question of univocality appears to be extended. In a move that seeks to both historicize Adorno's aporetic constructions and distance them from contemporary cultural criticism, his writings – especially his reflections on culture and art – are portrayed as the illustrations of an outmoded and exclusivist modernist worldview (Hohendahl 1995: 10). Formal experimentations and the ensuing complexities of Adorno's writings are viewed not as genuine attempts at resistance, but depicted as the forbidding elitist scripts seeking to deepen and perpetuate the distinctions between the 'high' pursuits of art and philosophy on the one hand and the 'low' culture of everyday life on the other (Jameson 1979). In such criticism, the tone of his writings – seen to be simultaneously despondent and authoritarian – is blamed for the supposed failure and irrelevance of his *Darstellungsformen*.

3.1.3 The Poststructuralist Appropriation

The third point of view, broadly categorized as poststructuralist, is less dismissive – and far more ambivalent – as regards Adorno's attempt to craft a philosophical discourse which recognizes the mediation of truth through language and seeks to theorise the aconceptual without reducing it to a set of principles (Hohendahl 1995: 7, 217). The poststructuralist appropriation of Adorno has as its primary foci his conception of negative dialectics and its unrelenting epistemological questioning, as also his incisive critique of the Enlightenment. According to this narrative, Adorno is construed "as a rigorous antimetaphysical thinker who struggle(d) against any form of (Hegelian) synthesis" (Hohendahl 1995: 7). Moreover, his assault in *Negative Dialectics* on the traditional epistemology of identity and the valorisation of the nonidentical are rendered as instances of proto-poststructuralism (Hohendahl 1995: 12). A more recent, but similar, approach employs a rereading of Adorno in order to reflect on the 'newer' questions raised by poststructuralist theory (Buhler 1999: 163). Such mappings delineate a pattern of shared philosophical affinities: contrapuntal readings of traditional philosophy, distrust of philosophical systems, critique of rationality, and the inclination towards the marginal. Thus, in this reinterpretation, poststructuralism and Adorno's philosophical writings are depicted as "parallel efforts to recover an ethics of alterity by way of an immanent overcoming of the tradition of philosophical idealism" (Pensky 1997: 6).

3.2 Limits of Adorno-Criticism

However, none of the major modes of English language Adorno-criticism engages with his theory and praxis of *Darstellungsformen* in an immanent manner. The post-Habermasian, Wellmerian and the postmodernist perspectives read his *oeuvre* selectively. Adorno's engagements with a variety of genres of presentation – from the aphorism, to the essay, and then to the radio talk – are levelled out into a unitary stylistic elaboration. Many of the postmodernist critiques of Adorno's style of writing appear to be at best only a partial assessment of his work, in that they do not take into account, for instance, his late and occasional writings, and his more accessible utterances on the questions surrounding postwar Germany and modern democracy (Pickford 1997; 1998). Furthermore, according to Hohendahl (1995: 12), "postmodernist discourse sets its own normative standards against which Adorno has to be read". Where discursive forms are concerned, it may be observed, that postmodernism advocates – and raises to quasi-normative status – a selection of formal strategies, tropes and tones, describing them as openended, intertextual, irreverent etc. Ironically, some other modes of expression, which do not correspond to this postcanonical canon, and that are reminiscent of an older, now much discredited modernism, are hastily rejected. Thus, as formal experimentation in Adorno is reduced to a mere set of techniques, the possibility of multivalent readings is denied to his writings, and his aphoristic, paratactic compositions are rendered as lifeless figures creating a fetish of difficulty.

3.3 The Poststructuralist Lure

The poststructuralist appropriation, on the other hand, is exceptionally alluring. Looked at in a more-or-less decontextualised manner, Adorno's negative forms, profoundly critical of conclusive philosophical edicts, appear to be compatible with the poststructuralist formulations which reject traditional philosophy's conceptions of truth. However, according to Hohendahl (1995: 217), the poststructuralist appropriation of Adorno's negative dialectic "neglects the specific constellation out of which his concept of truth emerged". Moreover, such attempts seem to be inattentive to those moments in Adorno's writings that seek to preserve the concept of totality and those that value conceptual rigour. Lastly, the anti-Humanism of much of poststructuralist thought is clearly at odds with the particular hermeneutic configuration within which Adorno states his philosophy.

3.4 Adorno Translated

Perhaps the most immanent, and therefore, productive contemporary relationship with Adorno's linguistic and stylistic forms of writing may be witnessed in some of his more recent English-language translations (For instance, Adorno 1991c; 1992; 1997b; 1998a, and Hullot-Kentor 1992). A close reading of Adorno-translations reveals that the conscious, self-reflexive, and to some extent, literal, renderings tend to be more coherent and comprehensible than the unselfconscious versions that try to put together a readable Adorno in English (Jameson 1990: ix-x; Hullot-Kentor 1985; 1989; 1992; Buhler 1999). It is evident that the project of translating Adorno's works calls for an understanding of the complex and diverse traditions of German Idealism, Critical Theory and philosophical aesthetics. However, above all, it demands a self-aware and critical engagement with the *Darstellungsformen* employed in his *oeuvre*, as also with the very task of translation.

3.5 Adorno's Anglophone Afterlife

Invoking Walter Benjamin's (1992) celebrated study of the task of the translator, one could claim that the English language Adorno-translations constitute his 'afterlife' in the Anglophone world. To be more precise, they comprise the posthistory of Adorno's exilic relationship with the English language and the Anglo-American culture (Adorno 1998b; 2002a; Hohendahl 1995: 21-44). Adorno's afterlife in English is not an evenly laid out, self-assured discourse. On the contrary, it is an intensely self-conscious and multivalent *oeuvre*, struggling to cope with the 'difficulty' of Adorno's utterances, beset with its lack vis-à-vis the 'original'. In other words, it is a body of writing not entirely at home with itself. Furthermore, Adorno-translations also depict a multilayered process where interpretive subjectivities wrestle with and perform the close connections between writing and politics. Thus, the 'difficulties' involved in the carriage of Adorno's complex and resistant forms into English could also be viewed as creative possibilities as much as simultaneous linguistic, intellectual and cultural anomalies, in that Adorno-translations traverse the narrow channels between understanding and confusion, familiarity and alterity. The narrative of Adorno's afterlife in English could be rendered in terms of distortion, appropriation, misinterpretation, and loss; however, it could also be retold as a story of deep intellectual engagement whose lived parallel may be witnessed in Adorno's experience as an émigré in the English-speaking countries (Claussen 2003).

3.6 Translation as Resistance

Thus, 'translation' emerges as a central trope (Niranjana 1995: 5) that may be employed to apprehend the dual moment of exile and resistance, in the context of Adorno's Anglophone reception. It could be suggested that the English language Adorno-translations be read as an emergent borderline discourse that seeks to translate what it posits as the 'untranslatable' (Weber 1997). In the process, it struggles to render into English forms of writing that may either remain utterly incomprehensible to the reader in the newer locations of reading, or redefine the very logic of reading in those contexts (Hullot-Kentor 1997: xi). Therefore, in the very act of translation, the *Problematik* of discursive resistance is taken onto a newer level, as Adorno's 'foreign' afterlife becomes a hybridised embodiment of diverse critical subjectivities belonging to different times and cultures. In this sense, the translation-endeavour could be seen as a profoundly political act, an inevitable and significant aspect of discursive resistance in our contemporary hermeneutic frames.

Works Cited

Adorno, T. W. (1993a; [1951]). *Minima Moralia: Reflexionen aus dem beschädigten Leben.* Frankfurt am Main: Suhrkamp.

Adorno, T. W. (1997a [1967]). *Prisms*. Tr. Samuel & Shierry Weber. Cambridge, Mass.: MIT Press.

Adorno, T. W. (1973). *Negative Dialectics*. Tr. E. B. Ashton. New York: Continuum.

Adorno, T. W. (1991a [1974]) *Drei Studien zu Hegel*. Ed. Gretel Adorno & Rolf Tiedemann. Frankfurt am Main: Suhrkamp.

Adorno, T. W. (1992 [1977]). *Prismen: Kulturkritik und Gesellschaft*. Frankfurt am Main: Suhrkamp.

Adorno, T. W. (2002a [1978]). *Minima Moralia: Reflections from Damaged Life*. Tr. E. F. N. Jephcott. London: Verso.

Adorno, T. W. (1991b [1981]). *Noten zur Literatur*. Frankfurt am Main: Suhrkamp.

Adorno, T. W. (1991c). *Notes To Literature, Volume One*. Tr. Shierry Weber Nicholsen. New York: Columbia UP.

Adorno, T. W. (1992). *Notes To Literature, Volume Two*. Tr. Shierry Weber Nicholsen. New York: Columbia UP.

Adorno, T. W. (1993b) *Hegel: Three Studies*. Tr. Samuel & Shierry Weber. Cambridge, Mass.: MIT Press.

Adorno, T. W. (1997b). *Aesthetic Theory*. Tr. Robert Hullot-Kentor. Minneapolis: University of Minnesota Press.

Adorno, T. W. (1998a). *Critical Models: Interventions and Catchwords*. Tr. Henry W. Pickford. New York: Columbia UP.

Adorno, T. W. (1998b). "Scientific Experiences of a European Scholar in America". Adorno 1998a, 215-42.

Adorno, T. W. (2001). *Negative Dialectics*. Tr. Dennis Redmond. This translation is available as a freeware at the URL: <http://www.efn.org/~dredmond/nd.html>.

Benjamin, Walter (1992 [1973]). "The Task of the Translator". *Illuminations*. Ed. Hannah Arendt, tr. Harry Zohn. London: Fontana, 70-82.

Buhler, James (1999). "Adorno Today". *Modernism/Modernity* 6.2, 161-68.

Claussen, Detlev (2003). "Intellectual Transfer. Theodor W. Adorno's American Experience". Unpublished paper. Presented at Max Mueller Bhavan, Mumbai.

Gibson, Nigel & Andrew Rubin, ed. (2002). *Adorno: A Critical Reader*. Oxford: Blackwell.

Habermas, Jürgen (1983a). *Philosophical-Political Profiles*. Tr. Frederick Lawrence. Cambridge, Mass.: MIT Press.

Habermas, Jürgen (1983b). "Theodor Adorno: The Primal History of Subjectivity - Self-Affirmation Gone Wild". Habermas 1983a, 99-109.

Habermas, Jürgen (1987a). *The Philosophical Discourse of Modernity: Twelve Lectures*. Tr. Frederick Lawrence. Cambridge: Polity Press.

Habermas, Jürgen (1987b). "The Entwinement of Myth and Enlightenment: Max Horkheimer and Theodor Adorno". Habermas 1987a, 106-30.

Hohendahl, Peter Uwe (1995). *Prismatic Thought: Theodor W. Adorno*. Lincoln: University of Nebraska Press.
Horkheimer, Max & T. W. Adorno (2002). *Dialectic of Enlightenment: Philosophical Fragments*. Tr. Edmund Jephcott. Ed. Gunzelin Schmid Noerr. Stanford, Cal.: Stanford UP.
Huhn, Tom, ed. (2004). *The Cambridge Companion to Adorno*. Cambridge: Cambridge UP.
Hullot-Kentor, Robert (1985). "Adorno's Aesthetic Theory: The Translation". *Telos* 65, 147-52.
Hullot-Kentor, Robert (1989). "Back to Adorno". *Telos* 81, 27-29.
Hullot-Kentor, Robert (1992). "Notes on *Dialectic of Enlightenment*: Translating the Odysseus Essay". *New German Critique* 56, 101-08; for his translation of the essay in the same issue, see "Odysseus or Myth and Enlightenment", 109-41.
Hullot-Kentor, Robert (1997). "Translator's Introduction". Adorno 1997b, xi-xxi.
Hullot-Kentor, Robert (2004). "Right Listening and a New Type of Human Being". Huhn, ed. 2004, 181-97.
Jameson, Fredric (1979). "Reification and Utopia in Mass Culture". *Social Text* 1.1, 130-48.
Jameson, Fredric (1990). *Late Marxism: Adorno, or, The Persistence of the Dialectic*. London: Verso.
Jay, Martin (1996). "Preface to the 1996 Edition". *The Dialectical Imagination: A History of the Frankfurt School and the Institute of Social Research, 1923-1950*. Berkeley: University of California Press; originally published (1973) Boston: Little Brown.
McCarthy, Thomas (1987). "Introduction". Habermas 1987a, vii-xvii.
Nicholsen, Shierry Weber (1991). "Translator's Preface". Adorno 1991c, ix-xi.
Nicholsen, Shierry Weber (1992). "Translator's Preface". Adorno 1992, ix-x.
Nicholsen, Shierry Weber (1997). *Exact Imagination, Late Work: On Adorno's Aesthetics*. Cambridge, Mass.: MIT Press.
Niranjana, Tejaswini (1995). *Siting Translation: History, Post-Structuralism and the Colonial Context*. Hyderabad: Orient Longman.
Pensky, Max, ed. (1997). *The Actuality of Adorno: Critical Essays on Adorno and the Postmodern*. Albany: State University of New York Press.
Pickford, Henry W. (1997). "Critical Models: Adorno's Theory and Practice of Cultural Criticism". *The Yale Journal of Criticism* 10.2, 247-270.
Pickford, Henry W. (1998). "Preface". Adorno 1998a, vii-xii.
Rubin, Andrew (2002). "The Adorno Files". Gibson & Rubin, ed. 2002, 172-90.
Said, Edward W. (2002). "Adorno as Lateness Itself". Gibson & Rubin, ed. 2002, 193-208.
Weber, Samuel (1997). "Translating the Untranslatable". Adorno 1997a, 10-15.
Weber, Samuel (2002). ""As though the end of the world had come and gone" or *Allemal ist nicht immergleich* – Critical Theory and the Task of Reading". Gibson & Rubin, ed. 2002, 379-99.
Wellmer, Albrecht (1991).*The Persistence of Modernity*. Cambridge, Mass.: MIT Press.
Weyand, Jan (2001). *Adornos Kritische Theorie des Subjekts*. Lüneberg: zu Klampen Verlag.

Leena Petersen

Poetics of the Intrinsic

On Walter Benjamin's Concept of Allegory

> For it pleases me, all for your sake, to row
> My own oars here on my own sea,
> And to soar heavenward by a strange avenue,
> Singing you the unsung praises of Death.
> (Pierre Ronsard) [1]

On the Concept of Allegory in General

By and large, the form of allegory has become a fundamental theme not only in literary studies. The concept of allegory is commonly understood by the definition of a story, a poem or a picture. The allegory might as well contain a hidden meaning. Etymologically, the term of allegory means to say something differently (compare: Duden 1997: 29). Furthermore, one distinguishes between allegory and allegorese. Allegory is defined as the representation of an abstract concept by a rational concrete image, often by the help of personification, whereas allegorese means the interpretation of a text which expounds a hidden meaning behind the sense of the word. Homer founded the first well-know allegorese in order to shape a justification towards philosophy (Meyers 1998: 203).

Within *Ästhetische Grundbegriffe*, allegory is characterised by metaphorising which is extended about a whole text or context, i.e. the rhetoric term of *metaphora continua*: here, a whole context becomes metaphorised within one and the same image sphere. Examplarily, the representation of a state or lifestyle through the image of shipping is considered as allegory. Vice versa, allegory can be defined as well as personification of an abstract concept. In contrast to the symbol, allegory and metaphor provide a functional linguistic coherence between primary and figurative sign (Haverkamp 2000: 55ff).

[1] From *Hymn de la Mort* by Pierre Ronsard, the very poem served as motto for Benjamin's convolute on Baudelaire in *The Arcades Project* (Benjamin -1940: 228).

According to the *dtv Lexikon der Antike*, the need to introduce the form of allegory comes into existence under the assumption of a high level of abstraction. Besides, a religious or public sense of mission is required, in which a striving for a universal artistic representation comes into terms with the world as its source. Such striving turns the whole cultural heritage into a kind of material of metaphorising. First, that material would be received by the help of allegorese and then, by means of an allegory, it would be moulded into the moving force of meaning of a new worldview (compare: von Albrecht 1965: 105).

Walter Benjamin's Allegorical Process as Model of Interpretation

Deducing from the latter, one can refer to Walter Benjamin's very complex concept of allegory – in case one reduces from that definition the implied personal medium of reflection within the use of allegory, and, instead, considers the opportunity of historical interpretation – which will be explained in the following.

In Fin de Siècle, Benjamin resumed the term of allegory within his intended postdoctoral thesis. In this work, *The Origin of the German Mourning Play*, he develops a definition of allegory which is characterised above all by its ambiguity and its regard towards past as images which are still existent in the present. That very image could be regarded as a 'death mask of history' – which will be explained in the following. Within his unfinished text of the *Arcades Project*, Benjamin applies the concept as dialectical picture puzzle as well as its conceptual continuation, the dialectical image, within, e.g., the elements of architecture, fashion, panorama, painting, modes of lighting, the mirror and the figure of the flâneur, particularly, in the context of his studies of Baudelaire.

Regarding the "pedagogic side of this undertaking" (of the *Arcades Project*), Benjamin quotes, correspondingly, Rudolf Borchardt: "To educate the image-making medium within us, raising it to a stereoscopic and dimensional seeing into the depths of historical shadows" (Benjamin 1940: 458). Following this quotation, one has not only to conclude that Benjamin's project deals with history but with the introduction of images into his proposed methodology. Arising from that very aim, one important element of his interpretative concept is the spatio-temporal form of allegory. By developing the concept of allegorical consideration, Benjamin intended hence a new form of culture critique and spatial-temporal historiography in opposition to a purely linear way of historical consideration.

The Benjamin researcher Howard Caygill explains regarding Benjamin's early method, that time and space are introduced as worldly categories of 'immanent critique'. These are developed in Benjamin's late work into terms of 'strategic critique'(Caygill 1998: 34f, 64ff), to be precise, into allegory and dialectical image as topologies of interpretation.

Conventionally, one considers space and time as part of sense-perception; hence the senses are often opposed to reason. In particular, Kant distinguished between concept and intuition (compare: Kant 1787: 168f), though he attempted to bring these polarities together under the auspices of time and space. The general rationalist distinction, however, raises the question of whether these categories are themselves rational concepts and, therefore, whether spatial-temporal perception should be regarded as primarily sensuous. Without asserting the entitlement to solve this issue here, it has to be stated, that the introduction of space and time as yardsticks of reality concerns an area between rationality, imagination and sense-perception. It concerns hence an intrinsic field in which these concepts are correlated. Regarding this intrinsic field, the question arises, how historiography might be possible on such a foundation. To solve this issue, another closer consideration upon the conception might be helpful.

As to the Kantian conception in *The Critique of Pure Reason*: here, Kant attempts to bring logos (the word) and aesthesis (the image) into line by means of schematism that accommodates the otherwise heterogeneous images and words into spatio-temporal rhythmic patterns (Kant 1787: 37ff). However, Kant states that one cannot find the essence in appearance (Kant 1787: 1ff). Here, his concept differs with Walter Benjamin's approach. Although, Benjamin takes up the topologies of time and space by means of frameworking the interpretation of reality, he regards semblance as a significant way of analysing reality's form of expression. That is to say: Benjamin regards truth and material as inseparable terms; therefore, the semblance of appearance which comes into existence in material and which forms the structure of reality, has to be examined.

For example, regarding the taste of beauty which comes to be relevant for the interpretation of an epoch and represents, consequently, an essence of that very time, Benjamin states within an earlier text:

> Beauty is not appearance, not a veil covering something else. It itself is not an appearing, but is essence throughout, one which, admittedly, remains essentially itself only when veiled. It may be that appearance is everywhere illusion, but the beautiful appearance is the veil for that which is most necessarily veiled. For neither the veil nor the veiled object is the beautiful but the object in its veil. If uncovered it would prove itself to be infinitely insignificant. (Benjamin 1922: 351)

This thesis can be understood on the assumption that the very mode of sense-perception within broad historical epochs relates itself to the change of the overall mode of being in the world of the 'historical collective' (compare: Benjamin 1940: 224).

Furthermore, one has to avoid the common misunderstanding of regarding Benjamin's work as pure interpretation of art. By introducing the image as an object of examination Benjamin intends by no means a new theory of art but is pointing both towards a new way of historiography and contemporary culture critique: namely, by using the semblance of an epoch to abandon it. Based on the assumption that reality, in general, is only given as fragmentary, history, consequently, decays – to be precise, into images, not stories (compare: Benjamin 1940: 476).

The concept of image, which is applied here, arises from the ambiguity of the Greek term *eikon*. According to Karl Kerényi, the image as such combines, for the moment, reality and deception. This very quality of reception refers back to the paradox features of pure surface and, at the same, depth of an image. If one would speak, nevertheless, of the depth of an image, one would have to bear in mind that the image which is admittedly sensuous as well as concrete existent would not possess this very depth but only in a transferred sense (compare: Kerényi 1964: 29). That very transferred sense will become important in the following.

However, even though the ambiguity of the imagistic is stated now as the one-dimensional appearance of which is often deceiving, there is no precise indication so far on which particular concept of image Benjamin's method is based. Despite the fact that Benjamin examines the work of art, defining the image as being purely based on art seems to be a by far too reduced definition, as already mentioned. Benjamin's approach deals not only with imagistic approaches for which language becomes the topos but with historiography. Following from this, one has to refer again to Kant's definition of sense-perception which describes the image as a product of the empirical ability of productive imaginative powers (compare: Kant 1787: 157ff).

This definition refers to a particular tendency in Benjamin's work where the empirical perception is connected with the imagistic: that very empirical ability expressed within the productive imaginative powers consists both of experience and imaginative potential. The potential of imaginative powers might lead to a memory of the past (which includes that very memory in form of images). Consequently, as both deceptive as well as true elements are inherent in the image as such, this very testimony of empirical reality can, obviously, only be found within the image. At the same time and correspondingly, the image can be characterised by magic and mythical elements – here, the attractive force of the imagistic and the unreal semblance dwell unalterably within the image.

Interpretation, therefore, is proposed to be relieved from such transfigured tendencies itself through the introduction of dialectics – nonetheless without giving up that very inclusion of the transfigured as, after all, the transfigured element has not only influence on the analysed time but consist of it. Here, linguistic dialectics supply the key-solution, namely, within language as form of interpretation, not within the image, as Benjamin states: „In the fields with which we are concerned, knowledge comes only in lightning flashes. The text is the long roll of thunder that follows" (Benjamin 1940: 456).

By reconstructing the linguistic illustration or rather 'physiognomic historiography', respectively – the imagistic expression of which comes to be starting point as well as principle of Benjamin's interpretation of history. Thereby, one has to bear in mind that its essence is constituted by Benjamin's concentration on capitalism's origin and development. This particular process is examined as it reveals, according to Benjamin, essential characteristics of the analysed epoch. Therefore, it has to be established that capitalism makes use of the image – in particular, of the image which one can find in the work of art and its continuations, e.g. advertisement. Hence, the development of the work of art, influenced by capitalism, demonstrates an important change concerning this period.

Even though the art critic Benjamin Buchloh sets forth Benjamin's concept of allegory as a way which would be of importance if one wanted to arrive at a more adequate reading of certain aspects of contemporary montage, its historical models, and the meaning of their transformations in contemporary art – and, therefore, remains within a purely art-historical interpretation –, his definition of allegory is noteworthy as he summarises the whole epoch of baroque under the headline of allegory. He defines Benjamin's approach in the regard of *The Origin of the German Mourning Play* as follows:

> Benjamin suggests that the rigid immanence of the Baroque – its worldly orientation – leads to the loss of an anticipatory, utopian sense of historical time and results in a static, almost spatially conceivable experience of time. The desire to act and produce, and the idea of political practice, recede behind a generally dominant attitude of melancholic contemplation. (Buchloh 1982: 44)

Here, the central quality of Baroque and its product, the allegorical contemplation, is seen in its spatial and temporal nature as well as its loss of the utopian sense of historical time. Besides, Buchloh's statement shares the opinion that Benjamin purely used this very approach of allegorical consideration as an instrument to describe the historical epoch he was working on. Unlike Benjamin's own intention, in Baroque, the idea of political practice happened to be 'receded behind the dominant attitude of melancholic contemplation'. The time which happens to be the central quality of the spatial form of allegory is adequately defined as standstill. This is the characteristic of the historical image as one can receive it from the present point of time.

Another conclusion can be drawn from this quotation: the 'almost spatially conceivable experience of time' marks the different meaning of experience. Without examining Benjamin's particular concept of experience in more detail, here, regarding the work of art and allegory, Howard Caygill states that

> the difficult relationship between critique and the philosophy of experience [...] manifests itself here in the torsion of space, time and the absolute. The exercise of critique indeed extends the concept of experience, pointing to new topologies of space, time and the absolute which are also new ways of being in the world. (Caygill 1998: 40)

Hence, this 'new way of being in the world' is characterised by a concentration of physiognomy phenomena.

Accordingly, Benjamin writes in the *Arcades Project*: "To write history means giving dates their physiognomy" (Benjamin 1940: 476)· This expression which describes Walter Benjamin's methodological course in one short sentence, underpins the need for the complex system of allegory as a counter-measure towards the totality of the pure symbol. As a reason for the use of the allegory, Benjamin offers the dialectical characteristic of the allegory in opposition to the one-dimensional symbol. Besides, the symbol is defined through its eternal nature and its mythical connotation. The expressive character of the allegory can be considered as its main feature and as well its attachment to the conventions of its time. Additionally, Benjamin describes the allegory rather as close to human existence because of its finiteness.

> Whereas in the symbol destruction is idealised and the transfigured face of nature is fleetingly revealed in the light of redemption, in allegory the observer is confronted with the *facies hippocratia* of history as a petrified, primordial landscape. (Benjamin 1928: 42)

Apologising the Artificial: The Allegorical Gaze of the Flâneur

In Walter Benjamin's method of allegorical consideration, one image, particularly, inhabits the allegorical procedure itself. Within the reception of the writings of Walter Benjamin, the figure of the flâneur has become extremely popular. However, this figure can only be defined according to his own awkwardness in contrast to conventional interpretations: Benjamin mentions the flâneur in one regard with trade and traffic as characteristics of the street, namely as commodity himself: "The flâneur sabotages the traffic. Moreover, he is no buyer. He is merchandise" (Benjamin 1940: 458). Thus, the figure of the flâneur is marked by his role within the transfigured moment of reality. Stating more precisely, Benjamin draws the subsequent conclusion:

> The flâneur is the observer of the marketplace. His knowledge is akin to the okkult science of industrial fluctuations. He is a spy for the capitalist, on assignment in the realm of consumers. (Benjamin 1940: 427)

According to Benjamin, the imaginativeness of Baudelaire has, by means of an example, its foundation within the melancholic figure of the flâneur. This phenomenon which is transfigured in itself becomes obvious in the way of looking at Paris:

> This poetry is no hymn to the homeland; rather, the gaze of the allegorist, as it falls on the city, is the gaze of the alienated man. It is the gaze of the flâneur, whose way of life still conceals behind a mitigating nimbus the coming desolation of the big-city dweller. (Benjamin 1940: 10)

The function of the figure consists, therefore, in its allegorical gaze which would be mainly signified by melancholy. Within *The Origin of the German Mourning Play,* Walter Benjamin defines the relationship between melancholy and visualisation – mourning and ostentation – among others, as caused by

> the self-absorption, to which these great constellations of the worldly chronicle seem but a game, which may, it is true, be worthy for attention for the meaning which can reliably be deciphered from it, but whose never-ending repetition secures the bleak rule of a melancholic distaste for life. (Benjamin 1928: 140)

Baroque melancholy is hence regarded as interdependent through its characteristic concentration on world events and its objects as a reaction to an attitude which is originally facing death. Therefore, melancholy is capable of intensifying itself: "Even from the heritage of the renaissance did this age derive material which could only deepen the contemplative paralysis" (Benjamin 1928: 140).

Here, within the central qualities of melancholy, one origin of Benjamin's methodological focus on the image as well as the way of contemplation is indicated. Later on, materialism will become a conceptual foundation for representing the allegorical "image of reality" (compare: Benjamin 1928: 44ff): accordingly, Howard Caygill states that Benjamin's examination of the Baroque drama investigates the culture of emerging capitalism while his *Arcades Project* considers the culture of high capitalism (compare: Caygill 1998: 57).

The mask-like revival of the world of Baroque which has lost its meaning plays hence an important role for melancholy as, how Benjamin sets forth, „mourning is the state of mind in which feeling revives the empty world in the form of a mask, and derives an enigmatic satisfaction in contemplating it" (Benjamin 1928: 139). However, melancholy's 'enigmatic satisfaction' within allegorical contemplation of worldly phenomena concentrates purely on its objects. Benjamin explains this occurrence as follows:

> Every feeling is bound to an *a priori* object and the representation of this object is its phenomenology. Accordingly the theory of mourning, which emerged unmistakably as a *pendant* to the theory of tragedy, can only be developed in the description of that world which is revealed under the gaze of the melancholy man. For feelings,

> however vague they may seem when perceived by the self, respond like a motorial reaction to a concretely structured world. (Benjamin 1928: 139)

Regarding the flâneur, the 'motorial reaction to a concretely structured world' consists in allegorical contemplation of spatio-temporal images as general attitude; correspondingly, Benjamin describes the allegorist's experience of flânerie as following:

> The 'colportage phenomenon of space' is the flâneur's basic experience. Inasmuch as this phenomenon also – from another angle – shows itself in the midnineteenth-century interior, it may not be amiss to suppose that the heyday of flânerie occurs in the same period. Thanks to this phenomenon, everything potentially taking place in this one single room is perceived simultaneously. The space winks at the flâneur. What do you think may have gone on here? (Benjamin 1940: 418f)

The form of allegory has, obviously, changed since Baroque, although, the concentration on outward appearances remains to be the main attitude of allegorical contemplation. However, the main object of the flâneur is, as mentioned before, the city (of Paris). On this matter, the arcades are regarded as being something in-between the street and interieur. If one would like to speak of a trick of physiologies, it would be the one proved by feuilleton: namely, to consider the boulevard as interieur. The street comes to be the flâneur's habitation. He is in-between facades at home like the bourgeois within his own four walls. The shining enamelled company's nameplates are as good and better as wall hanging decoration than the oil paintings within a drawing room. Walls would form the desk on which he would rest his notepad. Newsstands would become the flâneur's libraries as well as cafés happen to be the bay windows from which the flâneur would look down after finishing work. The fact that life in all its varieties would initially thrive between grey paving stones as well as the grey background of despotism – that would have become the political ulterior motive of the very literature to which the physiologies belonged to (compare: Benjamin 1938: 539).

The flâneur himself comes to be a commodity within his identity as bohemian:

> In the flâneur, the intelligentsia sets foot in the marketplace – ostensibly, to look around, but in truth to find a buyer. In this intermediate stage, in which it still has patrons but is already going to familiarize itself with the market, it appears as the bohéme. (Benjamin 1940: 10)

The indecisive economic attitude corresponds to the political function of the flâneur: initially, the flâneur was at home in the field of professional conspirators; after that, he sought his area of activity in the army and then within the lower middle class, occasionally, even in the proletariat. Whether Baudelaire sympathised with the clerical reactionary development or the rebellion of 1848 – according to Benjamin, Baudelaire's political insight remained to be immediately expressed. In this regard, a popular anecdote enlightens one revolutionary act of Baudelaire in which he was found with a rifle in the streets of Paris, shouting: "Down with General Aupick!" General Aupick happened to be Baudelaire's stepfather (Benjamin 1938: 515).

However, the *Communist Manifesto* brought the political existence of the bohemian to an end. Though, "Baudelaire's poetry draws its strength from the rebellious pathos of this class. He sides with the asocial. He realizes his only sexual communion with a whore" (Benjamin 1940: 10). How deeply rooted the dandy might be in his connection with the "asocial", can be diagnosed by an astonished Baudelaire: "Why don't the poor wear gloves when they beg? They would make a fortune" (Benjamin 1940: 235).

At the same time, the flâneur himself is characterised by indifference according to his own state of alienation. However, although he is looking for refuge in the crowd, the flâneur would be better defined as counter-example of throng as „the crowd is the veil through which the familiar city beckons to the flâneur as phantasmagoria" (Benjamin 1940: 10). Within the crowd, as a recorded expression of Baudelaire informs only a 'despicable blockhead' would be bored (compare: Benjamin 1938: 539). In this designed world, the department store comes to be a real treasure house of considerations to the flâneur: "There his fantasies were materialized. The flânerie that began as art of the private individual ends today as necessity for the masses" (Benjamin 1940: 895). Everything of the artistic existence of Baudelaire revolves around the artificial creation of his staged life. As a result of this, „the whole visible universe is but a storehouse of images and signs" (Benjamin 1940: 240), Baudelaire states in *L'Art romantique*. Within this department store of pictures and signs, the key to the emancipation from antiquity would be for Baudelaire allegorical interpretation.

Concerning Baudelaire, life and art are connected according to the concept of l'art pour l'art: Baudelaire speaks even about modern art as legendary translated form of outward life. That becomes obvious within the performed aestheticising of every day's life: vice versa, the consideration of art has an effect on the way of life which can be categorised as dandyism. The lifestyle which is in accordance with the dandy reflects – similar to a mirror – the art which portrays, in turn, the form of life. Baudelaire's idea of beauty, the contemporary taste of which might be revealing as it demonstrates the aesthetisising of every day's life.

On this matter, one can find Baudelaire's following expression of 1846:

> If ever your flâneur's curiosity has landed you in a street brawl, perhaps you will have felt the same delight as I have often felt to see a protector of the public's

> slumbers – a policeman or a municipal guard (the real army) – thumping a republican. And if so, like me, you will have said in your heart: 'Thump on, thump a little harder… The man whom you thumpest is an enemy of roses and perfumes, and maniac for utensils. He is the enemy of Watteau, the enemy of Raphael.'. (quoted in: Benjamin 1940: 263)

Unlike the republican, according to Baudelaire, the dandyish nature itself is comparable with a sunset or the decline of the morning star; as a means of identification, it would be full of glory, without heat and full of melancholy – "but alas, the rising tide of democracy [...] is daily overwhelming these last representatives of human pride" (Benjamin 1940: 239). This particular human pride could as well be considered as form of self-staging on the foundation of the pure surface or rather the superficial, respectively. As indicated by Benjamin, the artificial becomes so much more significant to Baudelaire as "nature [...] knows this one luxury: crime. Thus the significance of the artificial" (Benjamin 1940: 240). However, even though it suggests itself by now, Baudelaire's flânerie inhabits no moments of mythologising of his spatio-temporal perception as Benjamin emphasises: "It was owing to the genius of allegory that Baudelaire did not succumb to the abyss of myth that graped beneath his feet at every step" (Benjamin 1940: 268). Besides, it has to be emphasised that the latter enumeration of Baudelaire's ‚sadism' did by no means follow the intention of judging an individual poet of that very age. As Benjamin puts it, it

> is nothing more than his way of taking up the challenge which bourgeois society flings at the idle poet. This satanism is only a reasoned reprise of the cynical and destructive veilleities – delusions, in the main – that emanate from the lower depths of society. (Benjamin 1940: 379)

Following from the latter, the spatio-temporal quality of the flâneur's contemplation of the surrounding consists as well of a Baroque distaste for life. Although, the form of allegory has apparently transformed in modernity, Baudelaire, in particular, preferred the allegory as literary concept as – for him – it would provide the only solution for liberation from antiquity, as mentioned above. Baudelaire's existence circles around contemplating the world which is regarded as purely artificial – 'the whole visible universe becomes a storehouse of images and signs' – and, thereby, his allegorical concept includes, besides the spatio-temporal aspects, the main allegorical feature of finiteness: One can not only find a topography of the city of Paris in Baudelaire's poetry but – and that might decided Benjamin – his poetry shapes a conglomeration of modernity; the "'death-fraught idyll' of the city, however, is a social, a modern substrate. The modern is a principal accent of his poetry" (Benjamin 1940: 10). This "death-fraught idyll" of the city is signified by Benjamin as dialectical image and means, therefore, a historical image which is viewed from the present point. Accordingly, Theodor Adorno writes in a letter of 1935, describing Benjamin's concept – the definition of which Benjamin decided to quote:

> Dialectical images are constellated between alienated things and incoming and disappearing meaning. While things in appearance are awakened to what is newest, death transforms the meaning to what is most ancient. (Benjamin 1940: 466)

The Mirror as Paradigm of Dialectical Images

The appearance of the arcades which happen to be not only an object of flânerie might be defined through many elements: one fundamental phenomenon – as already mentioned – is the perception of the space which is characterised as ambiguous in itself. Benjamin writes:

> Ambiguity of the arcades as an ambiguity of *space*. [...] The outermost, merely quite peripheral aspect of the ambiguity of the arcades is provided by their abundance of mirrors, which fabulously amplifies the spaces and makes orientation more difficult. (Benjamin 1940: 877f)

According to Benjamin, he used the mirror as dialectical image as the mirror enables a reflection of the surrounding through its reproduction. The therefore simulated infiniteness of the arcades is extended as a moment of reflection. This constitutes the ambiguity of mirrors of this building as both the pretence as well as its dissolving may dominate the perception of the space. In this regard, the two-dimensional character of the mirror which suggests itself by now has, nevertheless, to be emphasised: in contrast to a representation in perspective, the mirror just reproduces. Correspondingly, Benjamin quotes a letter of Lahrs of 1837 the identity of which is unknown:

> Egoistic – that is what one becomes in Paris where you can hardly take a step without catching sight of your dearly beloved self. Mirror after mirror! In cafés and restaurants, in shops and stores, in haircutting salons and literary salons, in baths and everywhere, 'every inch a mirror'! (Benjamin 1940: 539)

As a result of its one-dimensional appearance, the mirror, strictly speaking, is not able to adopt a more differentiated attitude of consideration or rather representation.

At the same time, the extension of the space of the arcades illustrates the power of the architecture which is shown to the visitors of the building. Correspondingly, Benjamin describes its effect as follows: "Actually, in the arcades it is not a matter of illuminating the interior space, as in other forms of iron construction, but of damping the exterior space" (Benjamin 1940: 539).

Consequently, the remarkably character of the mirror is intended to simulate a horizon in which the mythologising of the world of material objects finds its place as well as the focus at the outward appearance of the human figure. The deception makes use of a connection of different spaces caused by reflection: mirrors which are put up on the walls of cafés take their surrounding – for example, the street – into its own space through the reproduction of it. Simultaneously, Benjamin states the connection of visitors of the arcades with commodity offered for sale through the reproduction of a mirror.

Benjamin's essay *The Work of Art in the Age of its Mechanical Reproduction* was first part of *The Arcades Project*. The topos of the contemporary reality of the 1930s which is represented in this text states the separation of the human from its own reflection caused by the technology of reproduction in the film. As a result, the flexibility of this fundamental element of economy is stated in which the human is, thereby, set back towards its outward appearance. The human astonishment in front of a camera would be the same as the surprise in front of a mirror by looking at the own appearance. However, as a result of the introduction of new technology, the own appearance has become removable from the own person (compare: Benjamin 1939).

The deceptive character of the mirror implies, consequently, through its widening within the technology of film the result to support the artificial construction of a representative character. The outstanding feature of the representative character consists of broadcasting in huge numbers. Regarding public spreading, Benjamin recurs on the development of fascism: the new media means there the possible broadcasting of a political agitator and, therefore, an extension of his sphere of influence. Concerning the new technologies and its connection with mythology, Benjamin states

> only a thoughtless observer can deny that correspondences come into play between the world of modern technology and the archaic symbol-world of mythology. Of course, initially the technologically new seems nothing more than that. [...] [However,] by the interest, it takes in technological phenomena, by the curiosity it displays before any sort of invention or machinery, every childhood binds the accomplishments of technology to the old worlds of symbol. (Benjamin 1940: 461)

In the beginning it was stated, that the introduction of space and time as yardsticks of reality concerns an area between rationality, imagination and sense-perception. It concerns hence an intrinsic space in which these concepts are correlated. According to Benjamin, his method of working emerges out of the realisation that only within the dialectical spatial image a presentation of the past in relation to the contemporary would be convertible:

> Image is dialectics at a standstill. For while the relation of the present to the past is a purely temporal, continuous one, the relation of what-has-been to the now is dialectical: is not progression but image, suddenly emergent. (Benjamin 1940: 462)

Regarding this, Benjamins's method of allegorical interpretation provides a seldom – unfortunately unfinished – methodological way of including the appearance of an epoch even though it is engaged with the character of semblance – or rather precisely for that reason, respectively.

In conclusion – regarding Benjamin's method of imagistic allegorical consideration – the question arises, of whether historiography and culture critique are possible without the use of images which inhabit the perception of time and space. Particularly, modernity implied the tendency of the imagistic. Therefore, it has to be examined if a presentation of that very epoch and its effects – i.e. historiography – can be translated into practice by ignoring establishing and revealing elements of non-conceptual images. However, by pointing towards the imagisitic, one has to bear in mind that the introduction of such elements holds the danger of conceptual adaptation to imagistic tendencies, i.e. accommodating mythical elements itself and of the very time which is examined. In the regard of such approaches – and the following quite revealing quotation builds the concluding consideration – Benjamin writes that the

> images are to be thought of entirely apart from the categories of the 'human sciences', from so-called habitus, from style, and the like. For the historical index of the images not only says that they belong to a particular time; it says, above all, that they attain to ligibility only at a particular time. And, indeed, this acceding 'to legibility' constitutes a specific critical point in the movement at their interior. Every present day is determined by the images that are synchronic with it: each 'now' is the now of a particular recognisability. In it, truth is charged to the bursting point with time. (This point of explosion, and nothing else, is the death of the *intentio*, which thus coincides with the birth of authentic historical time, the time of truth.) [...] The image that is read – which is to say, the image in the now of its recognisability – bears to the highest degree the imprint of the perilous critical moment on which all reading is founded. (Benjamin 1940: 462f)

Works Cited

Benjamin, Walter (1991 [1922]). "Goethes Wahlverwandtschaften". *Walter Benjamin: Gesammelte Schriften I.1.* Ed. Rolf Tiedemann. Frankfurt/Main: Suhrkamp.

Benjamin, Walter (1977 [1928]). *The Origin of the German Tragic Drama.* Tr. John Osborne. London: New Left Books.

Benjamin, Walter (1991 [1938]). "Das Paris des Second Empire bei Baudelaire". *Walter Benjamin: Gesammelte Schriften I.2.* Ed. Rolf Tiedemann. Frankfurt/Main: Suhrkamp.

Benjamin, Walter (1991 [1939]). "Das Kunstwerk im Zeitalter seiner technischen Reproduzierbarkeit". *Walter Benjamin: Gesammelte Schriften I.2.* Ed. Rolf Tiedemann. Frankfurt/Main: Suhrkamp.

Benjamin, Walter (1999 [1940]). *The Arcades Project: Prepared On the Basis of the German Volume Edited by Rolf Tiedemann.* Tr. Howard Eiland & Kevin McLaughlin. Cambridge & London: Belknap Harvard.

Buchloh, Benjamin H. D. (1982). "Allegorical Procedures: Appropriation and Montage in Contemporary Art". *Artforum* XXI.1, 43-56.

Caygill, Howard (1998). *Walter Benjamin: The Colour of Experience.* London: Routledge.

Duden (1997) (ed. by Annette Klosa, Werner Scholze-Stubenrecht & Matthias Wenke). *Duden. Das Deutsche Herkunftswörterbuch: Etymologie der deutschen Sprache.* Mannheim et al: Dudenverlag.

Haverkamp, Anselm (2000). "Allegorie als Figur". *Ästhetische Grundbegriffe Bd.1.* Ed. Karlheinz Barck, Martin Fontius, Dieter Schlenstedt, Burkhart Steinwachs & Friedrich Wolfzettel. Stuttgart & Weimar: Metzler. 55-70.

Kant, Immanuel (1787). *Critik der reinen Vernunft.* Riga: Johann Friedrich Hartknoch.

Kerényi, Karl (1964). *Griechische Grundbegriffe: Fragen und Antworten aus der heutigen Situation.* Zürich: Rhein-Verlag.

Meyers Lexikonredaktion (ed.) (1998). *Meyers Großes Taschenlexikon Bd.1.* Mannheim: Bibliographisches Institut.

von Albrecht, M. (1965). "Allegorie und Allegorese in der Literatur". *dtv Lexikon der Antike Bd.1: Philosophie Literatur Wissenschaft.* Zürich/Stuttgart: dtv, 103-106.

Rocío Orsi Portalo

Greek Tragedy and Political Agency

From Foucault's point of view, modern societies have a paradoxical structure: on the one hand, governments must protect values of vital importance for their population, such as freedom or security; but on the other hand, the means used to protect freedom and security seem to threaten just the same freedom and security that want to be protected. The Italian philosopher Roberto Esposito tries to fully explain this paradox by using the concept of "immunity".[1] Immunity, from the Latin *immunitas*, is a metaphor borrowed from the Biological Sciences, which evokes the common biological fact that every organism develops a protection system against the dangerous effects of invading microorganisms. Like an immunological system, every society develops different protection mechanisms with the aim of defending itself from all possible agents that could constitute a menace to it. These agents can be both internal and external enemies. Extrapolating this idea to modern societies, it is interesting to notice that we tend to use these immunity figures in a wide range of situations. For instance, the concept of preventive war, as it was used to justify the war against Iraq, involved the same kind of antinomic structure we are pointing out. The very concepts of law or right, as interpreted by Benjamin and Derrida, do not oppose to violence, they rather entail it.

Roberto Esposito realizes that it is very difficult to get out of this negative dialectic. In this paper I would like to develop Esposito's idea further applying his approach to a different, but related, field: that of the symbolic reproduction of social life and its relation to political agency. Here I suggest a possible way to respond to the difficulties of fighting violence in a non violent manner, i.e., violence can be combated by representing it and its causes, arguing that through its representation, its actualization can be prevented. I think we can find one clear example of this idea in one of the most important cultural institutions of the history of Ancient Greece: the Theater and, narrowly, the Tragedy. It seems to me that Athenian drama can be considered to function as an instrument of civic identity and, in a more positive sense, as a means of self-celebration or

[1] Esposito (2002).

self-affirmation of the polis; but it can also be considered as a tool for reflexive thinking and, in a more negative sense, as an instrument to accomplish a criticism of the existing values and conventions of the Athenian democracy. Taking all this into account, my aim will be to show how representation of violence can contribute to the elimination, or the catharsis, of actual violence in a real historical society.[2] Moreover, tragedy can be seen as one of the different devices invented by the Athenian society in order to provide itself with a space in which the expression of dissent was possible in a legitimate and safe way. In this sense, "acting" or "performing" violence in theater could be considered as a type of political action. Political agency in our present day representative democracies is located overall in political parties, but in Athens, where parties did not officially have a representative function, political agency was more dispersed and, as I hope to show in this paper, tragedy could be counted among the different forms in which it was manifested within Greek society. Of course, this device had – and it *must* have had – some limits in its working, limits self-imposed by democracy and directed to avoid the eventually self-destructive effects of excessive criticism. Therefore peace and security were ensured by the delicate, sometimes precarious, balance between the representation of violence and its concrete actualization.

During the Classic Period, to which the totality of the extant tragedies belong, Athenian citizens were deeply concerned with the problem of civil war (*stasis*). In the Archaic Period, the possibility of civil war was a real threat for both the rich and the poor people of Athens, and sometimes a popular tyranny could emerge against the abuses of the aristocracy. But these kind regimes were incompatible with the high sense of freedom housed by the Athenian citizenship. The establishment and consolidation of the Athenian democracy coincided with the Persian Wars, and therefore Athenians were aware of the perils surrounding the small Greek polis, but also depositaries of a sharp conscience of the importance of political freedom. In fact, political freedom was conceived as freedom from subordination to external rulers (i.e. freedom from Persian satraps), as well as freedom from subjection to internal rulers (i.e. democracy). The main threat to political freedom was civil war and, given the multiple factions into which the Athenian political arena was divided, its mere possibility constituted a really powerful threat.[3]

In Greek theater we can find a set of elements that, as a whole, contributed to maintain the characteristic plurality of the Athenian society and the sense of

---

[2] We can see in the homeopathic conception of *kátharsis* a precedent of the *immunitas* vocabulary. On catharsis as a term related to physicians' use of homeopathic remedies see the Oxford Edition of the *Poetics* by Lucas, Appendix II, "Pity, Fear and Katharsis".

[3] Thucydides draws a vivid picture of the extreme violence reached in a civil war in his description of Corcyra's civil war (*The Peloponnesian War*. III, 82-83).

unity that was so useful to preserve social stability and, consequently, political freedom. The festivals, as well as the plays performed during them, were a very important tool to produce civic cohesion and promote the self-celebration of the city. In order to explain how this collective self-affirmation of the city could be accomplished, we can mention some historical details. Firstly, performances took place in the context of public festivals (called "City Dionisias") which deeply contributed to the formation of a collective consciousness: there, people became aware of their political freedom as Athenian citizens by participating in the public sphere, and by this procedure the city itself became aware of its nature. Before entering the theater and, once inside the building but before the beginning of the play, Athenians performed a set of rituals such as processions, libations, honoring social benefactors, and finally a parade of war orphans; all these rituals emphasized social identity and cohesion. The dramatic performances themselves were collective rituals and, as such, contributed to the formation of a civic identity: in this sense, they worked as real elements of political agency.[4]

The whole city (citizens and government) was involved in the organization, funding, performance, and attendance of the dramatic festivals. The government – that is to say, citizens who temporally occupied the position of archons – chose poets, choirs and sponsors (*choregas*), and provided the prizes and salaries for poets and actors, whereas private sponsors provided the choirs' salaries and costumes. On the other hand, the upkeep and safety of the theaters was financed by the tickets. In fact, the total costs of the festivals were comparable to those of military campaigns.

Moreover, democratic life accustomed Athenians to be "audience", and therefore to play an active role in the collective sphere: war, banquets, courts, assemblies, sports or poetic contests were usual activities for every free citizen.[5] Indeed, during the dramatic spectacles the difference between audience (observers) and participants was progressively effaced. In their everyday political life, Athenians encountered speeches and activities very similar to those characteristic of theater, specially, because of their agonistic features.[6] In addition, before the construction of the theater buildings, there was no difference between the space of the Agora and that of the theater.[7] Consequently, theater was supposed to be an integral part of political or civic life and therefore a really important political factor.

---

[4] For classic approaches to the Greek theater production see Pickard-Cambridge (1968), Taplin (1978) and, more recently, Wiles (2000).

[5] Goldhill (1997: 54).

[6] Ober (1989).

[7] Wiles (2000, ch.1).

Since they were believed to perform a valuable civic function, actors, poets and chorographers had to be Athenian citizens, "in sharp contrast to the theatre in Rome, where acting was rather despised as something foreign, effeminate, fake, licentious, in short illegitimate and un-Roman".[8] The judges of the contests were selected among ten representatives of each tribe, through a public procedure and by lottery, the democratic formula par excellence. Participating in festivals was considered so important that during the Dionisias the Court and Assembly, the two most important organisms of political life, were suspended. Furthermore, the State ensured the attendance of the poor at the performance by providing them with the so-called "theoretic funds", i.e. the money equivalent to the price of the ticket.

Besides all this, the representation of the city's political system could be found in the very disposition of the audience seats, and we can also see the structure of the *polis* represented in the theater buildings: for instance, some seats were assigned to those who held public office, such as generals, archons, the Dionisos' priest or the members of the Council. Also, young men (*epheboi*) and war orphans were honored, and the ten tribes were seated separately. The treasure of the Delos League, symbol of the Athenian political and military superiority, was exposed in the theater during performances. This symbolic representation of the city was directed to the citizens themselves as well as to the guests attending the festivals: through it Athens deployed its social, cultural, political and military leadership for its own members and to foreigners, among them ambassadors, merchants or dignitaries from other cities.[9]

Furthermore, the ideological content of the dramatic plays contributed powerfully to the celebration of the city of Athens. Sometimes this goal was reached by the recurring use of panegyrics and oppositions between barbarians and Greeks or between Athenians and other Greeks, showing the supremacy of Athens over the entire world and the supremacy of citizens over non-citizens. In all the extant tragedies we find a pervasive presence of an ethnocentric, let us call it Athenian-centric, outlook that both produces and reproduces the moral, political, and social worlds of Athenians.

To sum up, Athenian dramatic festivals were occasions to show the national pride and promote social cohesion. This communal spirit supported by festivals was an indispensable tool to avoid the violence derived from the most fearful of evils, civil war. But, on the other hand, one can discern in the festivals and, particularly, in the plays represented there, a faithful illustration of the conflicts, problems, contradictions and paradoxes animating the political and social life of Athens. In short, it can be said that tragedy characterized an extremely useful

---

[8] Cartledge (1997: 18).

[9] See: Connor (1989) on the Athenian leadership and its relation to theater.

tool for critical thinking. The accomplishment of a deep critical self-exam in the context of Athens was made possible overall by the attainment of a distant gaze, and this distant gaze is precisely what the tragic performance contributed to create.

There is a wide range of theatrical or poetic mechanisms that help creating this necessary distance. First, tragedy represents general problems in a broad way. We can find some devices in the text, such as the constant use of traditional wisdom maxims (*gnomai*), which were very useful in order to create the required sense of generality in the formulation of the conflicts as conflicts that concerned everybody in the city.[10] Secondly, poets set the stories in different heterotopics, i.e. "other spaces" placed in a mythic past where the audience could hardly perceive any traits of their own world.[11] As a consequence of their archaic figuration, these spaces had an important charge of authority and idealization, which also contributed to create distance from the everyday world. For example, the city of Thebes functioned as an anti-Athens, always under the risk of civil war and disintegration, and in contrast to which Athens was configured as a place of integration and problem solution.[12] On the other hand, Argos usually appeared as a city/non-city, so to speak as a plastic city: sometimes it looked like Athens in its unanimity, sometimes like Thebes in its internal division.[13] But many other times, tragedy is set in more abstract and remoter scenarios, so distant from the ordinary life of the city, that they were often located even outside the boundaries of Greek civilization: Troy, Aulis, Tauride, Egypt, Caucasus Mountains, an isolated cave at Lemnos or a war camp. This was one of the most important devices directed to yield the critical distance required to accomplish a deep reflection on Athenian conventions and values: what people could hardly imagine to occur in Athens could easily be imagined to happen in Thebes, Argos, in Asia or on a lost desert island.

In addition, the characters of the tragedy were usually designed as people of noble linage, according to an aristocratic social frame that was obsolete in democratic Athens; this lack of realism in the delineation of characters placed them very far from the reality of Athenian citizens. The use of masks and costumes also helped to locate the action and characters in a space of symbols and fiction, and this symbolic universe was populated not by particular human beings but, as Aristotle put it, by general "types" of human beings.

The plots of the extant tragedies usually deployed the strains between community and the individual, or among different individuals, and there are

---

[10] On this topic see: de Romilly (1995: 46).

[11] *Heterotopics* are spaces from where it is possible to deconstruct the norms sustaining a certain cultural order. See: Foucault (1987).

[12] Zeitlin (1990: 144-50); Vernant/Vidal-Naquet (1989 ch. 8); Vidal-Naquet (2002: 74).

[13] Dupont (2001: 41).

many speech figures which contribute to accurately represent these conflicts. The same structure of tragedy favors the plurality of voices and the potential conflict involved in this plurality: by contrast to epic narrative or lyric poetry, tragedy shows the characters speaking by themselves in an immediate way and, in doing so, it is the voice of the characters themselves and not the poet's voice that is presented to the citizens. The voice of the poet hidden to the audience, poetic authority results undefined and it is distributed among all the characters, also those who, as women, barbarians and slaves, did not have any channel of public expression in real society. The polyphonic nature of tragedy enables it to express a plurality of conceptions and beliefs in the course of a single play, sustained by different characters, thus showing the irreducible plural nature of values and the impossibility of finding a straightforward solution to conflicts. In this sense the poetic world goes beyond the Athenian historical democracy breaking with its principle of exclusion, a key concept of political freedom. In fact, the flourishing of the Athenian democracy was based in a large extent on the work of slaves and the business of metics[14]: those who made the economical strength of Athens possible were not allowed to speak at the Assembly, but they did have voice in the theater.

Next, the very structure of the tragic plays shows a reflexive character. Indeed tragedies were made up of two alternating parts: first, we find the speaking parts performed by the main actors and representing individuals in the realization of different actions; on the other hand, we have the commentary on the actions by the choir, which represented the public voice or the outlook of the community. Therefore, at every moment the audience was confronted with different interpretations and reflections on what they were seeing, and by this they were always reminded of their being spectators. Literary figures which were pervasive in tragedy, such as irony, were very effective in yielding this kind of "spectatorial conscience"[15] by the temporal breach of fictional illusion. Finally, the fact that tragedies were written and not only recited is connected with their belonging to a transitional epoch from an oral to a literate culture. This fact is important in order to understand the tragedies' paradoxical position between conservatism and reflexive thinking, between its ludic spirit and its serious role. Tragedies were placed in the midway between the fully critical speech of written philosophy, such as the philosophy of Plato, and the traditional discourse represented by archaic poetry.[16] Therefore, by contrast to epic and

[14] See: Finley (1988).
[15] With this expression I hint at some parallels with the idea of "spectatorial vision", from which Adam Smith built his notion of the "Impartial Spectator", a key term relating to our concept of moral judgement.
[16] See Havelock on the critical role of written philosophy 1963. On Greek poetry and literacy see: Gentili (1984).

lyric, in tragedy the audience was not confronted with the exemplarity of the characters but with their essential ambiguity.[17] Here, tensions are supposed to be unavoidable facts of moral life.

As we have seen, tragedy was an instrument of self-celebration of the city and at the same time it was an essential resource for the critical review of moral and political values. This double-faced nature of tragedy can help to explain the great importance acquired by tragedy in the Athenian society as its deeply important dimension in the political action. As I have already mentioned, conflict was a reality in ancient Athens: many factions, parties and discussions constituted the matter of which everyday Athenian life was made up. Moreover, conflict was conceived as pregnant with violence. In peaceful civic life, conflicts, and thus violence, were actual and pervasive but inhibited. Representations of conflicts were, as the representation of violence, a way for people to think about them without the limitations imposed by the risk of their actual arising in real life. This freedom made it possible to express moral and political dissent in a safe and legitimate way. Therefore, tragedy accomplished the representation of conflicts among individuals, or between individuals and community. The outcome of these conflicts was, as we can see in the extant plays, extreme violence. The representation of these problems was fulfilled before the gaze of the audience, i.e. before the "civic gaze".[18] Finally, by this way Athenian society was enabled to review its own conventions and values critically and to weight up its potential conflicts.

As we can see, this critical exercise of reason protected Athenian society against itself: against its mistakes, its stubbornness, against its own betrayals. The representation of conflict and violence prevents the disastrous effects of a blind self-indulgence. However, one paradoxical consequence of the power of criticism of reason is that it can threaten the same social order that it is meant to protect. This menace is due to the violence contained in the very represented conflict, in the mere stating of irresoluble quarrels. In this sense, one possible byproduct of an exhaustive critical discourse could be the destruction of the essential values and the fall in a rough relativism. If tragedy provided the critical gaze required in order to avoid blindness and arrogant self-satisfaction, tragedy was at the same time, and by the same mechanism, a source of disorder and political rebellion. Therefore, self-criticism must have some limits.

The circumstances which made tragedy a form of self-celebration of the city set some limits to the critical function of tragedy. Only because there were limits tragedy could represent conflicts and violence and, by doing so, make people aware of the risks and dangers surrounding them. In fact, the Platonic approach

[17] See: Iriarte (1996: 8-12).
[18] As was stated by Goldhill (1997: 57).

to Greek theater shows how poets and judges used to flatter the audience[19]. We should not forget that tragedies were part of contests in which poets and *choregas* wanted to win, and for the sake of the contest one could not carry the criticism too far: some deep conventions had to be observed in every play. The tension between conventionality and criticism, between the reflexive review of one's conventions and the self-indulgent picture of oneself, makes this particular form of dissent possible in the frame of a society obsessed with achieving the maximum consensus. Therefore, plurality and unity were deployed in tragedy in a way which made them appear not as contradictory but complementary.

To conclude, in Athenian festivals, and especially in Athenian drama, people found themselves in front of a broken mirror[20] in whose surface they could look at themselves, reaching a deeper self-knowledge, even deeper than the one they could reach with their reflexive or philosophical resources alone, precisely because of this balanced tension between critical exam and self-indulgence.

Works Cited

Cartledge, Paul (1997). "Deep Plays: theatre as process in Greek civic life". Ed. Patrice Easterling. *The Cambridge Companion to Greek Tragedy*. Cambridge: Cambridge UP, 3-35.

Connor, Walter Robert (1995 [1989]). "City Dionysia and Athenian Democracy". *Classica et Mediaevalia* 40, 7-32.

De Romilly, Jacquelin (1995). *Tragédies grecques au fil des ans*. Paris: Les Belles Lettres.

Dupont, Florence (2000). *L'insignifiance tragic*. Paris: Gallimard.

Esposito, Roberto (2002). *Immunitas. Protezione e negazione della vita*. Torino: Einaudi.

Finley, Moses (1988). "The Fifth Century Athenian Empire: a Balance-Sheet", *Imperialism in the Ancient World*. Ed. P.D.A. Garnsey & C.R Whittaker. Cambridge: Cambridge UP, 103-26.

Foucault, Michel (1987). "Des Espaces Autres". *Revue d'Architecture*. Paris, 46-49.

Gentili, Bruno (1984). *Poesia e pubblico nella Grecia antica. Da Omero al V secolo*. Roma & Bari: Laterza.

Goldhill, Simon (1997). "The Audience of Athenian Tragedy". *The Cambridge Companion to Greek Tragedy*. Ed. Patrice Easterling. Cambridge: Cambridge UP, 54-68.

Havelock, Eric (1963). *Preface to Plato*. Cambridge, MA: Harvard UP.

Iriarte, Ana (1996). *Democracia y tragedia: la era de Pericles*. Madrid: Akal.

Ober, Josiah (1989). *Mass and Elite in Democratic Athens. Rhetoric, Ideology, and the Power of the People*. Princeton: Princeton UP.

Lucas, D.W. (1968). "Appendix II. Pity, Fear and Katharsis". *Aristotle's Poetics*. Oxford: Oxford UP, 273-290.

Pickard-Cambridge, Arthur (1968). *The Dramatic Festivals of Athens*. Oxford: Oxford UP.

---

[19] See: *Gorgias*, 501e-502d, *Republic*, 492b-c and overall *Laws*, 658d-559d and 817a-e.

[20] As it was established by the French scholar Vidal-Naquet (2002).

Platón (1999). *Las Leyes.* Intro, translation & notes: Francisco Lisi. Madrid: Gredos.
Taplin, Oliver (1978). *Greek Tragedy in Action.* Oxford: Routledge.
Thucydides (1960). *The Peloponnesian War.* Tr. Richard Crawley; Ed. Sir Richard Livingstone. New York. Oxford UP.
Vernant, Jean-Pierre & Pierre Vidal-Naquet (1989). *Mito y tragedia en la Grecia antigua, vol.II.* Madrid: Taurus.
Vidal-Naquet, Pierre (2002). *Le miroir brisé. Tragédie athénienne et politique.* Paris: Les Belles Lettres.
Wiles, David (2000). *Greek Theatre Performance: An Introduction.* Cambridge, MA: Cambridge UP.
Zeitlin, Froma (1990). "Playing the Other: Theater, Theatricality, and the Feminine in Greek Drama". *Nothing to do with Dionysos? Athenian Drama and its Social Context.* Ed. Zeitlin, Froma & John Winkler. Princeton: Princeton UP.

Ulf Schulenberg

Democratic Space and Poetic Agency

Emerson, Whitman, and the (Im)Possibility of Dissent

The notions of liberal democracy, liberal individualism, and, above all, liberal consensus are indispensable if one wants to conceptually grasp the ideological and rhetorical complexity of the U.S. (see, for instance, Hartz 1955). One might feel tempted to advance the argument that the U.S. appears more and more like a total liberal democratic space that at least pretends to be governed by a sort of consensus which leaves no room anymore for voices of resistance, heterotopian desires, and rhetorical forms of dissent. That this development has gained additional vehemence since 9/11 goes without saying. In this context one ought to see, however, that it was Alexis de Tocqueville who, in *Democracy in America* (1835, 1840), already called attention to the danger of an intolerance of heterodoxy in the U.S. Liberal theorists, from John Locke and John Stuart Mill to Richard Rorty, have always been enormously important for the development of the Left since they have been strong and creative proponents of the idea of liberal democracy and of the ideology of the free market, and thus critics of the notion of radical and fundamental social change. Leftist theorists, in other words, by criticizing and dialectically using the insights and suggestions of liberals, have been offered the possibility of clarifying their own position and of illuminating the contours of their program.

Since the early 1990s, no liberal has underlined the alleged triumph of liberal democracy more decidedly than Francis Fukuyama. In his neoliberal manifesto *The End of History and the Last Man*, Fukuyama maintains that liberal democracy "may constitute the 'end point of mankind's ideological evolution' and the 'final form of human government,' and as such constitute[s] the 'end of history'" (1992: xi). Moreover, he argues that liberal principles in economics have spread, and that the free market has "succeeded in producing unprecedented levels of material prosperity, both in industrially developed countries and in countries that had been, at the close of World War II, part of the impoverished Third World" (1992: xiii). Fukuyama's political and economic liberalism, this very strange mixture of neo-Hegelianism, Christianity, and distortions of Nietzsche and Kojève, is clearly teleological and governed by the notion of clôture. Richard Rorty, seemingly in agreement with Fukuyama's main suggestions, contends

> that no more romantic prospect stretches before the Left than an attempt to create bourgeois democratic welfare states and to equalize life chances among the citizens of those states by redistributing the surplus produced by market economies (Rorty 1998: 229).

Undoubtedly, there have been numerous attempts to attack or confront what I have termed the total liberal democratic space. As far as political activism is concerned, one only has to think of the various, partly very successful anti-globalization movements (see Schulenberg 2003a). With regard to leftist theory and the problem of space, a line of theorizing reaches from the Jamesonian idea of cognitive mapping to Hardt and Negri's materialist mapping of the new global world order in *Empire and Multitude*.

Apart from political activism and leftist theorizing, there is another form of resistance which has hitherto been neglected and to which I thus want to call attention. This is what I call poetic agency. In this article, I wish to discuss the question of whether poetic agency, that is, the rhetorical work of the strong poet in the Bloomian and Rortyan sense, can effect forms of genuine change in the confrontation with the omnipresent ideology of American liberal democracy. I shall start my argument with a discussion of Rorty's conception of a literary or poeticized culture. After that I shall seek to elucidate a main part of the history of the American strong poet by focusing on two authors of the American Renaissance: Ralph Waldo Emerson and Walt Whitman. What do these authors write concerning the role of the American poet? It is not only the aforementioned question of whether the poet is given the possibility of uttering genuine dissent and resistance which is of primary concern here. It will be argued that it is of utmost importance to realize that even the strong poet is in constant danger of eventually ending up as part of the powerful liberal consensus which has granted him a place on the margin of society. To put this somewhat differently, it is crucial to grasp that the total space of American liberal democracy does not necessarily seek to repress or marginalize radical energies and gestures of resistance, but that it also attempts to creatively redirect those energies and gestures so that they will eventually, and paradoxically, strengthen that which they originally opposed and fought against.

What this seems to boil down to is the rather bleak suggestion that not even poetic agency, the work of the strong poet as romantic radical redescriber, creative innovator, or inventor of new sets of metaphors, is capable of effectively confronting the hegemony of liberal democracy in the U.S. The liberal consensus with its subtle instruments of liberal cooptation not only emphasizes that socialism, and maybe even radical democracy, is incompatible with the ideology and rhetoric of America, but it also forces one to understand that critique of its premises and theoretical foundations can hardly ever become a fundamental challenge to its existence since that critique will be redirected so that its position will always already be within the space of liberal democracy, and it will therefore not be given the possibility of demarcating a genuine outside. However, one might argue that it is precisely the power of liberal ideology, seemingly omnipresent and insurmountable, that urges leftist theoreticians to relentlessly pursue the goal of establishing counterhegemonic sites of discourse. It will be demonstrated that the creative and innovative work of the American strong poet, in spite of all the severe difficulties and obstacles it has to confront, can play a crucial role in the creation of the possibility of counterhegemony.

The idea of a post-Philosophical culture has preoccupied Richard Rorty since his introduction to *The Linguistic Turn* (1967). It was central to many of the essays collected in *Consequences of Pragmatism* (1982), and it played a decisive role in the last chapter ("Philosophy Without Mirrors") of *Philosophy and the Mirror of Nature* (1979). In its most fully developed form the idea of a post-Philosophical culture as poeticized culture is one of the primary aspects of Rorty's *Contingency, Irony, and Solidarity*. The beginnings of this kind of culture can be detected at the end of the eighteenth century with the emergence of new vocabularies, yet it has not yet been fully realized. A liberal poeticized culture in its fully realized form would be antifoundationalist, antiessentialist, nominalist, fallible, and historicist through and through. Such a literary or poeticized culture urges us to grasp that only strong poets and Rortyan liberal ironists fully recognize the importance of contingency and the power of radical redescription. The most provocative, innovative, and stimulating heroes of a literary culture are strong poets in the sense of people who constantly long to redescribe things in new ways, who use words as they have never been used before, and who desire to expand the power of the human imagination. The strong poet, in other words, makes us realize the importance of self-creation, self-fashioning, and redescription. The strong poet as the creator of a new vocabulary can be P.B. Shelley elaborating on the power of "the unacknowledged legislators of the world" (*A Defence of Poetry*), or it might be Hegel moving at the limits of syntax and letting us realize that it is perfectly legitimate to advance the argument that sublation is but a form of innovative redescription. It is interesting to see that in Rorty's description of his ideal poeticized culture most of the crucial elements of his neopragmatist thinking come together: his antifoundationalism and antiessentialism, David-sonian and Wittgensteinian nominalism, Hegelian historicism, Darwinian naturalism, Nietzschean and Proustian perspectivism, as well as his Freudian conception of the human self (see: Schulenberg 2003b).

In one of his latest pieces, "Philosophy as a Transitional Genre", Rorty states a thesis which he has been repeating since the 1970s: "It is that the intellectuals of the West have, since the Renaissance, progressed through three stages: they have hoped for redemption first from God, then from philosophy, and now from literature" (2004: 8). For our purposes, the following question is of utmost interest: How does Rorty define the members of a literary culture, the literary intellectuals? His understanding of the function of the literary intellectual combines a Bloomian interpretation of the autonomy of the self with Emersonian self-reliance. A literary intellectual has constant doubts about the (final) vocabulary she is currently using, she does not want to get stuck in it. She longs to become acquainted with other ways of speaking, other ways of interpreting the purpose of life. For that reason she reads as many books as possible. By becoming acquainted with so many alternative vocabularies and ways of being human, the literary intellectual enlarges her self. Because of her reading she is introduced to a great number of alternative purposes, and ways of expressing those purposes, and she is thus given the possibility of radically questioning traditional vocabularies and explanations. To put it simply, the literary intellectual's reading leads to her self-creation, it offers her the possibility of creating an autonomous self. Rorty apparently agrees with Harold Bloom that the more books you have read, the more descriptions and redescriptions you have come across, the more human and at the same time autonomous you become (see Bloom 2000). A Rortyan and Bloomian autonomous self puts a premium on the attempt to creatively expand the present limits of the human imagination and it also seeks to demonstrate that the development from religion (God) to philosophy (Truth) to literature (novelty, imagination, redescription) is a story of increasing self-reliance.

It is crucial to understand that in a poeticized culture literary criticism is the presiding intellectual discipline. Liberal ironists think of literary critics as moral advisers because the latter have read many books and are therefore acquainted with many different vocabularies. Moreover, literary critics can offer new perspectives since they are often capable of bringing together books and authors which have hitherto been considered as incompatible. The critics' syntheses of, for instance, Nietzsche and Mill or Marx and Baudelaire or Sartre and Proust are examples of the power of creative redescriptions. Yet Rorty's contention is that the idea of ironist literary intellectuals governing a liberal poeticized culture and engaging in literary criticism in the broadest sense has to be problematized insofar as there is still a gap between those intellectuals and the public. They belong to a kind of avant-garde because they have the leisure and the money to spend most of their time reading books and redescribing persons and things. It is their job to convince the public, which consists mostly of metaphysicians, of the advantages of a nominalist and historicist literary culture. Rorty writes:

> The rise of literary criticism to preeminence within the high culture of the democracies – its gradual and only semiconscious assumption of the cultural role once claimed (successively) by religion, science, and philosophy – has paralleled the rise in the proportion of ironists to metaphysicians among the intellectuals. This has widened the gap between the intellectuals and the public. For metaphysics is woven into the public rhetoric of modern liberal societies. (Rorty 1989: 82)

The story I have been telling so far is one of increasing self-reliance, that is, it is an Emersonian story about the strong poet's desire for self-creation in a liberal poeticized culture. However, one ought to see that this is not only a story that centers on self-reliance and self-creation but also on fear. One might even say that it begins with fear. Rorty apparently follows Bloom in arguing that poetry begins in fear. The profound fear of the poet is to awake one day and find himself in a world he has not made, an inherited world. To put it another way, the poet fears to end his days in a world to which he has contributed nothing of distinction and which he has not creatively changed. The story of the strong poet's self-reliance begins with his "anxiety of influence," that is to say with his "horror of finding himself to be only a copy or replica" (Bloom 1973: 80). The poet fears that his most important creations do not really belong to him, but that they are the results of unconscious impressions which a stronger, more radical and impressive poet has left on his mind. Continually anxious about these impressions and their destructive effects on his work, the strong poet has to decide that he wants to get rid of them and that he does not want to be influenced by other poets. He wants to create his own impressions and leave his own mark on the world. He desires to leave his impressions on others and to redescribe the idiosyncracies of his individual self in a way that makes it attractive to others.

The strong poet, fearing to be regarded as a mere copy or replica, eventually finds out that by describing himself in his own terms he creates himself. Confronting his own con-tingency, acknowledging and appropriating it, is synonymous with the creative invention of a new language or a new set of metaphors. Following Rorty, the final victory of poetry in its ancient quarrel with philosophy might also be interpreted as the final victory of idiosyncratic metaphors of self-creation over metaphors of discovery. Instead of the will to truth which governed a philosophical culture, a post-Philosophical or literary culture would be dominated by the Nietzschean will to self-overcoming and self-creation. It should have become clear from what I have said so far that the Rortyan literary culture is a liberal (utopian) space which defines radical change almost exclusively as radical and innovative redescription and/as self-creation.

The Emersonian poet also does not accept, and does not need, someone else's description of himself. He is the self-reliant namer of things, the inventor of a new language, the destroyer of traditional forms, the teller of new experiences; he is representative of man and at the same time a liberating god. The poet, it seems, is the exact opposite of the "[m]eek young men [who] grow up in libraries" (1837: 56) of whom Emerson speaks in "The American Scholar." Like the American scholar, the Emersonian strong poet does not accept dogmas, traditional book-knowledge and education, the security of the past, popular opinion, systems of knowledge, and conformity. Furthermore, the scholar and the strong poet are both creative inventors of new languages and sets of metaphors, both are self-reliant redescribers who long for self-creation, self-trust, and self-overcoming. Both, moreover, want to leave their impressions on others and thereby change their minds. In the Rortyan sense they aim to redescribe things and persons, and by doing so they wish to convince or persuade others to adopt this new vocabulary. As Emerson says of the American scholar's power of redescription: "Not he is great who can alter matter, but he who can alter my state of mind. They are the kings of the world who give the color of their present thought to all nature and all art, and persuade men by the cheerful serenity of their carrying the matter, that this thing which they do is the apple which the ages have desired to pluck, now at last ripe, and inviting nations to the harvest" (1837: 65). While the idea of "giv[ing] the color of their present thought to all nature and all art" is the typical gesture of the Emersonian, Nietzschean, and Bloomian strong poet who desires to be more than a mere replica, the notion of persuasion, which is of utmost importance here, indicates that the strong poet is also always a subtle and elegant rhetorician. No one has taught us more vehemently that rhetoric goes all the way down, and that there is no outside to rhetoric, than Stanley Fish (cf. Fish 1989). However, in contrast to Fish's radically rhetoricized antifoundationalism, as a form of neopragmatism, the work of the Emersonian scholar as/and poet intends to have consequences in the world of practice. Emerson underscores in "The American Scholar" that the scholar, as a self-reliant innovator, starts from the rich world of practice and also ends there. Radically rejecting the courtly muses of Europe seems to be a truly worldly affair, as it were.

According to Emerson, the true poet "stands among partial men for the complete man" (1844: 242). Moreover, he "traverses the whole scale of experience, and is representative of man" (1844: 243). The true poet, as Emerson sees him, is "the sayer, the namer, and represents beauty. He is a sovereign and stands on the centre" (1844: 244). In contrast to traditional and conventional poets, for whom "the finish of the verses is primary" (1844: 245) and for whom meter and form are of utmost importance, the original poet is supposed to forever change our perspective by articulating a new and passionate thought, by singing a radically new experience. By constructing his own mind, the strong poet creates his own language, his own idiosyncratic set of metaphors. He does not care, or rather pretends not to care, about the vocabularies other human beings have left behind. The strong poet, in his Emersonian version, sings about his own time and place in his own words, and he waits for his readers to follow him along hitherto untrodden paths:

> The poet has a new thought; he has a whole new experience to unfold; he will tell us how it was with him, and all men will be the richer in his fortune. For the experience of each new age requires a new confession, and the world seems always waiting for its poet (Emerson 1844: 245).

In "The Poet", Emerson repeatedly underlines the power of the true poet as far as his use of language is concerned. Not only does the poet make the old use of symbols forgotten, or fills those symbols with new life, but he is also capable of stepping nearer to things than his fellow human beings. Since he can see the higher form within the form everyone else perceives, he can represent and communicate with the ideal. In other words, the poet can depict the essence of the higher form by means of his innovative and creative language. The point here is not that in some of his essays Emerson's idealism is grounded in an appearance-reality/essence distinction, a distinction which Rorty, for instance, has been severely criticizing since his early texts, but rather Emerson's suggestion that the poet's "speech flows with the flowing of nature" (1844: 252). The Emersonian poet sings the real nature of things, and he expresses the divine aura which shows itself in the aforementioned higher forms. To Emerson,

> the poet is the Namer or Language-maker, naming things sometimes after their appearance, sometimes after their essence, and giving to every one its own name and not another's, thereby rejoicing the intellect, which delights in detachment or boundary (Emerson 1844: 252).

In order to get close to the real nature of things, the poet as language-maker, this Adam-like figure or American Adam, must recognize "that he speaks adequately then only when he speaks somewhat wildly" (Emerson 1844: 255).

In view of what I have said so far the question arises as to what Emerson's own status is. Is he, who always abhorred the idea of a system of thought and who appreciated contingency, experiment, inconsistency, change, and transition, a philosopher? Or ought one to see him, who wrote only a few good poems, rather as a poet? A reading that concentrates on Emerson's antifoundationalism and pragmatism would surely tend to consider him a poet. In this context, John Dewey contends:

> Perhaps those are nearer right, however, who deny that Emerson is a philosopher, because he is more than a philosopher. He would work, he says, by art, not by

> metaphysics, finding truth 'in the sonnet and the play.' 'I am,' to quote him again, 'in all my theories, ethics and politics, a poet'; and we may, I think, safely take his word for it that he meant to be a maker rather than a reflector. His own preference was to be ranked with the seers rather than with the reasoners of the race, for he says, 'I think that philosophy is still rude and elementary; it will one day be taught by poets'. (Dewey 1929: 25)

This cannot be the place to discuss this crucial question in detail, yet I wish to call attention to Cornel West's suggestion that the Emersonian evasion of philosophy "views poetry and philosophy neither as identical nor as antagonistic but as different metaphor-deploying activities to achieve – by means of agon and struggle – specific aims" (1989: 73). Emerson in this sense is a strong poet as mediator, a mediator in a strictly non-Hegelian sense who does not seek to achieve a synthesis of poetry and philosophy but who rather elegantly uses different vocabularies for different purposes. It could be argued, to follow West's line of thought, that Emerson's primary concern is the dialectical tension, the play, between poetry and philosophy. His is a non-teleological and non-systematic way of speaking, which sometimes seems almost antimetaphysical and which illustrates a strong desire to make it new. Yet Emerson's newness, in spite of his severe critique of "the barbarism and materialism of the times" (1844: 261) and although he presented himself as a social reformer who fought against slavery and for women's rights (cf. Robinson 2004), must not be interpreted as offering a possibility of demarcating an outside to the total liberal democratic space. The Emersonian strong poet is an integral part of the American liberal consensus; a welcome voice of dissent (read: reformist desire) which eventually confirms and strengthens the rhetorical power of the hegemonic consensus. The Americanist Sacvan Bercovitch formulates as follows in this context:

> Even at the height of his heresy, Emerson's cultural roots were too deep for him to envision 'a new form' in the 'Art of Living' that would be anything other than a purified version of free enterprise [. . .] or a pastoral dream of agrarian laissez-faire (Bercovitch 1993: 325-26).

In his critique of Bercovitch's reading of Emerson, John Carlos Rowe goes even further when he contends

> that the 'liberal dissent' Bercovitch reads in Emerson's canon is not the energetic 'paradox' of these political writings, but rather the fundamental contra-diction beyond which Emerson cannot take his transcendentalism (Bercovitch 1997: 24).

I think that Rowe can be agreed with when he argues that "transcendental and practical politics continue to collide in his essays from 1844 to 1862, demonstrating the inappropriateness of Emerson-ianism for any practical social reforms" (1997: 24).

In many respects Walt Whitman was the paradigmatic namer, the sayer, the American bard who sang his beloved home country. Furthermore, in many respects Whitman is the paradigmatic strong poet who illustrates the possibilities and the power of self-creation, of a self-reliant individual inventing him- or herself. It should have become clear from what I have said so far that the Romanticism which unites Emerson, Whitman, and Rorty is not only grounded in the idea that the poets – as redescribers – are the true legislators of the world, but that it can also be detected in the strong emphasis on the importance of (Nietzschean) self-creation and contingency. The genuinely protean self of Whitman's "Song of Myself," for instance, continually rewrites and reinvents itself. This roaming, straying, and loafing self experiences the world in all its complexity and overwhelming simplicity, goes from situation to situation, person to person, landscape to landscape, and every new impression heightens its desire to remake itself. The self, the soul never far removed from the body, embraces contingency and contradiction and longs to bring a new form of life into being just by means of its creative and innovative use of words. America is in permanent flux, or so Whitman seems to hold, and the songs of the romantic radical redescriber are supposed to shape that which is yet to come. To Whitman, the poet as radical redescriber and prophet of continual self-invention is always a poet of democracy. While Whitman often changed his ways of expressing his democratic sympathies, from his expansive vision of democracy and his notion of populist democracy in the 1840s and 1850s to the decidedly gloomier picture he painted in the 1870s (e.g., in *Democratic Vistas*), he always made unequivocally clear that the crucial ideas of a democratic personality and a democratic soul could only be realized in the United States. Although Whitman, from the 1840s to the 1880s, was torn between his peculiar kind of liberalism and his understanding of radical democracy, his major texts, and this is no minor aspect, illustrate that Romanticism might also signify that the gesture of radical redescription productively comes together with the program of radical democracy. *Pace* Rorty, redescription does not necessarily go hand in hand with liberalism. However, the question arises as to whether Whitman's texts really are a powerful means in the confrontation with the power of the total liberal democratic space. Are they helpful as far as the development of counterhegemonic sites of discourse is concerned?

It is very interesting to see that Whitman's "Preface" to the first edition of *Leaves of Grass* (1855) contains some passages which could almost have been written by Rorty in his description of the characteristics and advantages of a literary culture. One parallel that ought to be mentioned here concerns the question of contingency. Just like the Rortyan strong poet or literary intellectual or liberal ironist, the Whitmanian great poet is not afraid of contingency. He handles it well: "He consumes an eternal passion and is indifferent which chance happens and which possible contingency of fortune or misfortune and persuades daily and hourly his delicious pay" (1855: 12). It is undoubtedly true that one might argue that there are sentences in the "Preface" which show that Whitman's poet is still a metaphysician or Platonist who is incapable of leaving behind foundationalist epistemology and essentialist thinking. Whitman maintains, for instance: "The poets of the kosmos advance through all interpositions and coverings and turmoils and stratagems to first principles" (1855: 19). In addition, one sometimes gets the impression that his statements are at least partly shaped by a mimetic understanding of art. Yet one should also recognize that a sentence like "Whatever satisfies the soul is truth" (1855: 23) is pragmatist and antifoundationalist all the way down. This understanding of truth is almost Jamesian in the radical nature of its critique of traditional definitions of this concept.

The role Whitman attributes to literature as far as the strengthening of American democracy is concerned cannot be overestimated. His ideal society seems to be a (secular-ized) literary culture governed by self-reliant strong poets, "the great literatus[es] of the modern" (Whitman 1871: 321) who illuminate that social and religious problems and ills can be treated by literature and who also elucidate the fact that American literature shapes American moral identity and moral consciousness. Regarding the power of literature, Whitman underscores that it "penetrates all, gives hue to all, shapes aggregates and individuals, and, after subtle ways, with irresistible power, constructs, sustains, demolishes at will" (Whitman 1871: 321). He moreover points out that

> in the civilization of today it is undeniable that, over all the arts, literature dominates, serves beyond all – shapes the character of church and school – or, at any rate, is capable of doing so. Including the literature of science, its scope is indeed unparalleled (Whitman 1871: 322-23).

Whitman's comments on the contemporary American situation in *Democratic Vistas* are truly severe. Although he has certainly not lost his faith in the future, *Democratic Vistas* is a partly bitter text. The American poet whose powers he joyfully praised in the first preface has not arrived, his own book has turned out to be a commercial failure, and the political situation of the U.S. seems truly bleak to the poet of democracy. To a certain degree, and this is a decisive change in comparison with his earlier texts, he even seems to have lost faith in the common people, as when he speaks of "the people's crudeness, vice, caprices" or warns against "the appalling dangers of universal suffrage in the United States" (Whitman 1871: 318). The impact of the Civil War, the direction of Reconstruction, and the spread of political corruption – these are only the main aspects that have darkened his outlook. With regard to the future of the U.S., Whitman poses the following question:

> [T]he problem of the future of America is in certain respects as dark as it is vast. Pride, competition, segregation, vicious wilfulness, and license beyond example, brood already upon us. Unwieldy and immense, who shall hold in behemoth? who bridle leviathan? Flaunt it as we choose, athwart and over the roads of our progress loom huge uncertainty, and dreadful, threatening gloom (Whitman 1871: 377).

Compared to his earlier prose texts, Whitman's stress in *Democratic Vistas* on the future of America ("Vistas"), the idea of democracy, and the notion of a genuinely American literature appears to be even stronger. Concerning the function of literature, his contention is that "the only reliable identity [is] the moral and artistic one," and he moreover underscores that his belief in "a new and greater literatus order – its possibility (nay certainty), underlies these entire speculations – and that the rest, the other parts, as superstructures, are all founded upon it" (Whitman 1871: 324, 379).

If one remembers the Bloomian suggestion that the more books you have read, the more human and autonomous you become, as well as Rorty's idea that the literary intellectual seeks to enlarge her self by becoming acquainted with many other alternative (final) vocabularies and ways of self-fashioning, then it becomes apparent that the Whitmanian kind of literary or poeticized culture in many respects is similar to the ones described by Bloom and Rorty. For instance, Whitman comments on the role of literature for the building of personality, that is, for the process of self-creation, as follows: "The literature, songs, aesthetics, etc., of a country are of importance principally because they furnish the materials and suggestions of personality for the women and men of that country, and enforce them in a thousand effective ways" (ibid.: 346-47). Simply put, it is the strong and romantic belief in the power of literature that links Emerson, Whitman, Bloom, and Rorty. These writers are united by their belief in what I have called poetic agency. The notion of radical or populist democracy as developed by Whitman in the first version of *Leaves of Grass* and other early texts can be considered as offering angles from which to criticize U.S. liberal democracy. Yet while the Whitmanian gesture was more radical than those underlying Emerson's and Rorty's texts, even he was incapable of completely escaping from the total liberal democratic space (in this context, see Cmiel 2000).

Nonetheless, one ought to realize that Whitman's texts are particularly valuable and useful since they show that it is legitimate to advance the argument that the strong poet might also be a poet of radical democracy. The Whitmanian early texts let us glimpse the rhetorical power of the creative and innovative strong poet in the confrontation with the ills and shortcomings of liberal democracy. They illuminate, even if only vaguely, that the notion of counterhegemony must not necessarily remain a utopian desire. However, there is one last point to which I would like to call attention. We have seen that in a literary culture the reliance on God, or Truth, is replaced by self-reliance. This American religion is a form of American exceptionalism that in many respects is highly problematic; its coordinates are an extreme idealism, a shallow liberalism and reformism, and partly even a petit bourgeois conservatism. The question whose importance I wish to underscore in this context is whether the notion of a literary culture also leaves room for the proposal that one might imagine the strong poet and radical redescriber as a *theorist* whose work has effects and consequences in the public sphere. In other words, my discussion of the power of poetic agency and of the (im)possibility of dissent in a total liberal democratic space has prepared the ground, as it were, for the analysis of another crucial problem. Can the (American) theorist, one might ask, find a position from which to radically critique liberal democracy, and can he or she at the same time avoid the danger of succumbing to another version of American religion? In the context of my argument this boils down to the question of whether theory can offer a special kind of redemption in a literary culture. A kind of redemption, that is, which differs from the one offered by literature.

Works Cited

Bercovitch, Sacvan (1993). *The Rites of Assent: Transformations in the Symbolic Construction of America.* New York: Routledge.

Bloom, Harold (1973). *The Anxiety of Influence: A Theory of Poetry.* New York: Oxford UP.

Bloom, Harold (2000). *How to Read and Why.* New York: Simon & Schuster.

Cmiel, Kenneth (2000). "Whitman the Democrat". *A Historical Guide to Walt Whitman.* Ed. David S. Reynolds. New York: Oxford UP, 205-233.

Dewey, John (1978 [1929]). "Ralph Waldo Emerson". *Emerson: A Collection of Critical Essays.* Ed. Milton R. Konvitz & Stephen E. Whicher. Westport, CT: Greenwood Press, 24-30.

Emerson, Ralph Waldo (1981 [1837]). "The American Scholar". *The Portable Emerson.* Ed. Carl Bode. New York: Penguin, 51-71.

Emerson, Ralph Waldo (1981 [1841]). "Self-Reliance". *The Portable Emerson.* Ed. Carl Bode. New York: Penguin, 138-164.

Emerson, Ralph Waldo (1981 [1844]). "The Poet". *The Portable Emerson.* Ed. Carl Bode. New York: Penguin, 241-265.

Fish, Stanley (1989). "Rhetoric". *Doing What Comes Naturally: Change, Rhetoric, and the Practice of Theory in Literary and Legal Studies.* Durham: Duke UP, 471-502.

Fukuyama, Francis (1992). *The End of History and the Last Man.* New York: Avon Books.

Hartz, Louis (1955). *The Liberal Tradition in America: An Interpretation of American Political Thought Since the Revolution.* New York: Harcourt, Brace & World.

Robinson, David M. (Ed.) (2004). *The Political Emerson: Essential Writings on Politics and Social Reform*. Boston: Beacon Press.

Rorty, Richard (1989). *Contingency, Irony, and Solidarity*. Cambridge: Cambridge UP.

Rorty, Richard (1998). *Truth and Progress: Philosophical Papers*, Vol. 3. Cambridge: Cambridge UP.

Rorty, Richard (2004). "Philosophy as a Transitional Genre". *Pragmatism, Critique, Judgment: Essays for Richard J. Bernstein*. Ed. Seyla Benhabib and Nancy Fraser. Cambridge: MIT Press, 3-28.

Rowe, John Carlos (1997). At *Emerson's Tomb: The Politics of Classic American Literature*. New York: Columbia UP.

Schulenberg, Ulf (2003a). "Gestures of Despair and Gestures of Hope – Remarks on Globalization, Neoliberalism, and the Specters of Theory". *Americanization – Globalization – Education*. Ed. Gerhard Bach, Sabine Broeck, and Ulf Schulenberg. Heidelberg: Universitätsverlag Winter, 1-33.

Schulenberg, Ulf (2003b). "Wanting Lovers Rather than Knowers – Richard Rorty's Neopragmatism". Amerikastudien/American Studies, 48.4., 577-608.

Shelley, Percy Bysshe (1993 [1840]). *A Defence of Poetry, The Norton Anthology of English Literature*. Sixth Edition, Vol. 2. Ed. M.H. Abrams. New York: Norton, 752-765.

West, Cornel (1989). *The American Evasion of Philosophy: A Genealogy of Pragmatism*. London: Macmillan.

Whitman, Walt (1977 [1855]). "Preface to Leaves of Grass". *The Portable Walt Whitman*. Ed. Mark Van Doren. New York: Penguin, 5-27.

Whitman, Walt (1977 [1871]). Democratic Vistas. *The Portable Walt Whitman*. Ed. Mark Van Doren. New York: Penguin, 317-382.

agency and its spatio-temporal dimensions

Holger Rossow

Time-Space Compression, Flexible Accumulation and the Problem of Political Agency in the Age of Globalisation

Introduction

> Capital can always move away to more peaceful sites if the engagement with 'otherness' requires a costly application of force or tiresome negotiations. No need to engage, if avoidance will do. (Bauman 1998: 11)

The emergence of the system of 'flexible accumulation' (Harvey 2001) can be interpreted as a response to the crisis of the Fordist mode of accumulation and has been closely interwoven with those processes commonly described as globalisation. Among many other developments, it has entailed a new phase of time-space compression, i.e. a shortening of time and a shrinking of space. Time-space compression has undoubtedly been a real process, but what needs to be challenged are claims that it will produce a homogeneous global social, economic, cultural or political space and the assumption that it is the decisive or even the only cause for the current changes. What further needs to be asked is how the general conditions, limits and opportunities of political agency have been redefined in a world in which according to Zygmunt Bauman the "technological annulment of temporal/spatial distances tends to polarize [the human condition]" (Bauman 1998: 18) rather than to homogenise it, a world in which capital continues to dominate not least because of its command over space and time. A world in which the physical presence of the top managers, let alone the owners, of capital is no longer essential and in which the consequences of increasingly flexible modes of accumulation undermine traditional forms of political agency.

Any attempt to claim or re-claim political agency – especially for those who are disadvantageously affected by certain aspects of the flexible regimes of accumulation in particular and globalisation in general – necessitates on the theoretical level to inquire into the causal nexus between flexible modes of accumulation and newly emerging patterns of the distribution of power and political agency. Apart from the avoidance of economic determinism, such an inquiry further requires to embed the current changes into a longer-term historical framework in order to work out not only discontinuities but also continuities between Fordist and post-Fordist regimes of accumulation and the respective patterns of distribution of power and political agency.

The first part of the paper sketches in the theoretical assumptions underlying the analysis and focuses in particular on the suitability of neo-Gramscian approaches. Sections two and three are devoted to a brief discussion of the continuities and discontinuities between the Fordist and the post-Fordist era. The problem of time-space compression, as a significant but by no means the only aspect of the move towards more flexible modes of accumulation and intensified processes of globalisation since the 1970s, is discussed in section four. Against this backdrop, the final section offers a preliminary analysis of the new conditions, limits and opportunities of agency on four distinct but interrelated levels at the beginning of the twenty-first century.

Theoretical approach

The following analysis is based on a number of theoretical assumptions which cannot be thoroughly discussed here, but should at least be indicated. The most general is that the character of the change towards more flexible regimes of accumulation, embedded in a multi-faceted process of globalisation, requires an approach that goes beyond the confines of traditional disciplines. The extremely complex relationship between the character of globalisation and disciplinary structures is convincingly put by Hay and Marsh, who propose an "inter-disciplinary, indeed post-disciplinary approach", when they claim that

> an account which privileges (empirically or, worse still, causally) one 'moment' (whether social, cultural, political or economic) can but only fail to capture the complexity and contingency of contemporary change. (Hay and Marsh 2000: 3)

Bruff reminds us in this context that "[w]e should [...] take seriously the possibility that globalisation as a discursive construct has real, causal effects on policy outcomes *and* on the process of globalisation itself" (Bruff 2005: 268). This position is clearly supported by Hay and Rosamond when they write that globalisation "has become a key referent of contemporary political discourse and, increasingly, a lens through which policy-makers view the context in which they find themselves" (Hay and Rosamond 2002: 148). They are also convinced that the discourse of globalisation "is likely to affect significantly political dynamics" (ibid.).

An inter-disciplinary approach to globalisation in general and flexible accumulation in particular accompanied by an acknowledgement of the crucial role of the *discourse* of globalisation is not only theoretically relevant but foregrounds the possibility of agency and highlights the potential role of the subject in the political process. Proceeding from this starting point and to avoid the trap of economic determinism a neo-Gramscian approach seems to be suitable to analyse the complexities of the move towards more flexible modes of accumulation and its global context. This is because, writes Woods, Neo-Gramscians "see the world economy as an overarching structure of knowledge, ideas and institutions which reflects the interests of dominant actors and within which competition takes place" (Woods 2001: 289). This structure itself is considered to be vital to understand the identities and preferences of the actors. At the same time, neo-Gramscians do not lose sight of the economic aspects because they consider capitalist competition as "the key driving force of the system [...] which is then constrained by the need of the powerful to gain the consent of the less powerful" (Woods 2001: 289). The latter point is echoed by Hall, who rejects the notion of "globalization as a non-contradictory, uncontested space", when he writes that "in order to maintain its global position, capital has had to negotiate and by negotiate I mean it had to incorporate and partly reflect the differences it was trying to overcome" (Hall 1991: 32). The same theme is taken up by Bruff when he argues that

> the hegemonic project of the transnational historical bloc [...] must articulate its vision in variable ways across different spatial and temporal horizons. The intention is to enable other social groups to 'consent' to the project via attaching meanings to globalisation that are not antagonistic to the meanings attached to globalisation by the transnational historical bloc. (Bruff 2005: 275)

Employing a neo-Gramscian approach, the question can be posed more adequately how certain preferences of actors develop, in which ways they reflect 'interest' and how these processes relate to politics, culture, the law and knowledge in general – not just economics or rational choice preferences. For Woods it is clear that there are no objectively definable 'interests' but that rather "the way actors understand their own preferences will depend heavily upon prevailing beliefs and patterns of thinking in the world economy, many of which are embodied in institutions" (Woods 2001: 288).

In view of the significance of economic aspects for the following analysis a brief discussion of the assumed relationship between economic and non-economic issues seems to be advisable. The usage of the terms 'Fordism', 'post-

Fordism' and 'flexible accumulation' might be interpreted as a return to simplistic base-superstructure conceptualisations. Hall in his reply to such criticism has pointed out that Gramsci used the term 'Fordism' "to connote a whole shift in capitalist civilization" (Hall 1996b: 225). For Hall – and this view is shared by the author – the term 'post-Fordism' does not assume any prior determining position of the economy, but

> it does insist – as all but the most extreme theorists and culturalists must recognize – that shifts of this order in economic life must be taken seriously in any analysis of our present circumstances. (Hall 1996b: 225)

Elsewhere Hall discusses the problem of 'economic determinism' with particular reference to ideology. Part of his more complex argument is that the "economic *cannot* effect a final closure on the domain of ideology, in the strict sense of always guaranteeing a result" (Hall 1996a: 44). There are two related reasons for this. The first is that "ideological categories are developed, generated and transformed according to their own laws of development and evolution; though, of course, they are generated *out* of given materials" (Hall 1996a: 44-5). The second reason is the "necessary 'openness' of historical development to practice and struggle" (ibid.: 45). The two reasons which Hall advances are equally relevant for our argument. The former allows us to analyse, for example, the *relationship* between flexible modes of accumulation and the neo-liberal discourse without assuming any direct *causality*. The latter is significant because unless we insist on this 'openness', there would be no room for agency and no room for change unless the economic sphere is radically remodelled – or in extremis, capitalism is abolished as an *economic* system.

The theoretical positions sketched in above are drawn from diverse sources but they achieve some 'pragmatic' coherence because they enable both an analytical and a practical approach that preserves at least some room for agency and choice, and does not – self-defeatingly – assume that 'neutral', 'objective', 'natural' or 'unavoidable' 'economic forces' or the 'logic of the market' necessarily lead to certain outcomes.

Historical background: Fordism

In order to bring out more clearly the character of the changes which have taken place over the last three decades the following section briefly looks at the preceding period. The avoidance of the fashionable phenomenon of historical amnesia (Hall 1991: 20) helps bring to the fore the new conditions at the beginning of the twenty-first century, the new players, the playing fields and the new rules but also the historical, structural and discursive continuities between the Fordist and the post-Fordist period.

Henry Ford believed, according to Harvey, "that the new kind of society could be built simply through the proper application of corporate power" (Harvey 2001: 126). The term Fordism later came to signal the ability to produce ever-larger quantities of standardised products for mass markets which in turn were consumed by workers with sufficient incomes. The emphasis on technological innovation (the introduction of the moving assembly line) and mass production sometimes obscures the more important reasons that led to the triumphant advance of the Fordist model of production in both capitalist and socialist economies:

> Fordism owed its success not to its capacity to produce and market goods on a wide scale but to its social and political consequences. It was an extraordinarily effective means both of controlling the labour process and of satisfying workers' aspirations at a material level. In the terms of a well-known formulation it turned proletarians into instrumental workers. (Waters 1995: 80)

Among the central objectives of Fordism in capitalist economies was the avoidance of the overaccumulative potential of capitalism, i.e. enhanced productivity and industrial production were to be met by increased consumer demand.

In contrast to the dramatic situation during the great depression in the 1930s, which for many had seriously questioned the viability of capitalism, the post-war period in the industrialised countries of the West was marked by a great deal of economic, social and political stability. However, the inherent contradictions of the capitalist system and the systemic conflicts of interest between capital and labour had not been solved but only temporarily contained. They became exposed again in the late 1960s and early 1970s against the backdrop of completed post-war recoveries of Western European and Japanese economies, excess production capacities, inflation, saturated national markets in most industrialised countries and under pressures of increasing international competition which resulted from the search for export opportunities. These trends forced many corporations "into a period of rationalization, restructuring, and intensification of labour control (if they could overcome or bypass union power)" (Harvey 2001: 145).

The post-war boom, according to Harvey, "was built upon a certain set of labour control practices, technological mixes, consumption habits, and configurations of political-economic power, and [...] this configuration can reasonably be called Fordist-Keynesian" (ibid.: 124). But he also argues that the period from 1965 to 1973

> was one in which the inability of Fordism and Keynesianism to contain the inherent contradictions of capitalism became more and more apparent. On the surface, these difficulties could best be captured by one word: rigidity. (ibid.: 141-2)

One way to confront these 'rigidities' seemed to be a more flexible regime of accumulation which is characterised in the following section.

Flexible regimes of accumulation since the early 1970s

Although there seems to exist a widespread acceptance that far-reaching changes have taken place since the 1970s, there is much less agreement concerning the particular character, range and depth of those changes, and how they could be theorised most adequately. Hall, in his analysis of the relationship between 'New Times' and Thatcherism, emphasised the transitional character of the changes in the 1970s and 80s and came to the conclusion that we are

> certainly not debating an *epochal* shift, of the order of the famous transition from feudalism to capitalism. But we have had other transitions from one regime of accumulation to another within capitalism, whose impact has been extraordinarily wide-ranging (Hall 1996b: 232)

Harvey was similarly concerned with the transitional character of the changes and argued that there "is always a danger of confusing the transitory and the ephemeral with more fundamental transformations in political-economic life" (Harvey 2001: 124). But he was convinced that changes of political-economic practices had occurred that were "sufficiently strong to make the hypothesis of a shift from Fordism to what might be called a 'flexible' regime of accumulation" (ibid.).

'Flexibility' has become one of the buzzwords in the globalisation discourses since the 1980s, a tenet of neo-liberalism, a constitutive part of political rhetoric, something that employers and employees equally need – but what does it really mean? For Dahrendorf, the word "is intended to convey something desirable, though for many it describes the price they have to pay [...] the word has so many connotations that it is hard to pin down to any particular meaning" (Dahrendorf 1996: 25). Among the many possible meanings of the term 'flexibility' are certainly the following: deregulation, less government interference, liberalisation, reduced labour protection, more part-time and limited-term employment, sub-contracting, lower taxes for companies, flexible investment and marketing strategies, increased downward pressure on wages and benefits, demands for higher 'mobility' of workers, i.e. to change employers, change locations of employment, to accept de-skilling and re-skilling, the readiness of employers and employees to respond to intensified rates of commercial, technological, and organisational innovation, etc. However, the changes do affect employers and employees, capital and labour in radically different ways. Bauman argues that

> the idea of 'flexibility' hides its nature of social relation: the fact that it demands redistribution of power, and implies an intention to expropriate the power of resistance of those whose 'rigidity' is about to be overcome. [...] 'Flexibility' only pretends to be a 'universal principle' of economic sanity, one that applies in equal measure to both the demand and the supply side of the labour market. The sameness of the term conceals its sharply different substance on each side of the divide. (Bauman 1998: 104)

The more immediate results of these changes so far include a radical restructuring of the global economy, a period of deindustrialisation in the USA, Japan and Western Europe and corporations that were taking advantage of more fluid and liberalised financial markets, cheaper labour, weaker regulations, and hence lower production costs through the relocation of manufacturing jobs to Third World countries and, after the disappearance of the Iron Curtain, also Eastern Europe. The opening up of Eastern Europe represented a particular challenge to workers in the industrialised countries of the West because it not only provided, especially for Western European companies, vast new pools of cheap, well-trained and docile labour but also governments which were devoted to neo-liberal policies that would have been difficult if not impossible to implement in the West. In the wake of this global restructuring of the international division of labour, territorial competition not only among different nations, but also regions and local communities for new investments or the maintenance of existing levels of investment has increased dramatically. This increased competition has often resulted in a 'race to the bottom', i.e. a lowering of standards and wages, reduced labour protection, fewer benefits, a downsizing of the number of full-time employees, subcontracted workers with lower wages and fewer benefits, etc., which currently also affects the core capitalist countries ever more strongly. Production processes are increasingly decentralised, i.e. the final products are composed of numerous parts which may be produced around the world, partly also due to reduced costs of transport and communication. The location of individual production sites and conditions of investment and production have become 'negotiable' – often dependent on the level of subsidies, the generosity of tax reliefs for potential investors, environmental standards, wage levels, health standards and the degree of labour protection.

In view of the problem of agency, the consequences of the shift towards more 'flexible' modes of accumulation must not be limited, however, to the technological changes nor to the modification of economic structures and organisational patterns. What is arguably more important in this respect are the political, cultural and social implications of post-Fordism. For Hall the concept does also signal "greater social fragmentation and pluralism, the weakening of older collective solidarities and block identities and the emergence of new identities as well as the maximization of individual choice through personal consumption" (Hall 1996b: 225).

Time-space compression as one dimension of flexible accumulation

The emergence of more flexible modes of accumulation has coincided with (but not been caused by) a new phase of time-space compression, the *material* base of which has mainly been provided by technologically driven changes: globally connected computerised communication networks, global electronic financial systems (with floating exchange rates) and global transport systems that have reduced the significance of time and geographical space. This, however, is not a novel phenomenon – and should therefore not be overestimated – because throughout the history of capitalism innovations have been dedicated to the removal of spatial barriers, to the overcoming of the constraints of space and time. Marx commented on this inherent characteristic of capitalism already in 1857 when he wrote that

> while capital must on one side strive to tear down every spatial barrier to intercourse, i.e. to exchange, and conquer the whole earth for its market, it strives on the other side to annihilate this space with time, i.e. to reduce to a minimum the time spent in motion from one place to another. The more developed the capital, therefore, the more extensive the market over which it circulates, which forms the spatial orbit of its circulation, the more does it strive simultaneously for an even greater extension of the market and for greater annihilation of space by time. (Marx 1981: 539)

The general effect of capitalist modernisation has been "very much about speed-up and acceleration in the pace of economic processes and, hence, social life" (Harvey 2001: 230). Waters shares that view and moreover assumes that flexible accumulation is a particular form of time-space compression:

> It was principally directed at reductions in turnover time [...], by the development of outsourcing, 'just-in-time' inventory systems, and small batch production. In its turn consumption patterns have experienced a similar temporal compression. (Waters 1995:57)

One dimension of time-space compression which is possibly more significant than the decreased relevance of time and space as a result of technological progress *per se* are its repercussions on the human condition. Ideas that globalisation and time-space compression will create one global space in which all of us can equally benefit from the new conditions in which distance and geographical borders increasingly lose most of their relevance simply do no stand up to closer scrutiny. Empirical evidence rather points to a situation which is simultaneously characterised by processes of globalising and localising. The two processes "sharply differentiate the existential conditions of whole populations and of various segments of each one of the populations" (Bauman 1998: 2).

Closely related to the effects of time-space compression outlined above is the question how the 'locals' can respond to global forces that seem to be able to penetrate borders easily, to cover distances, to reach into the most remote places of the world, to eclipse time and to a world "in which the 'globals' set the tone and compose the rules of the life-game" (ibid.). The following section widens the scope of the analysis again and discusses some of the new conditions, limits and opportunities of agency under conditions of globalisation, flexible accumulation and time-space compression.

New conditions, limits and opportunities of agency

The analysis of the new conditions, limits and opportunities of agency in an increasingly but unevenly globalised world presupposes simultaneously an acknowledgement of the global repercussions of many of the changes that have taken place over the last three decades and the necessity to avoid totalising assessments that gloss over the differences between different places and between different groups within particular countries. This means that despite the fact that "we are all being globalized [...] globalization must never be read as a simple process of cultural homogenization; it is always an articulation of the local, of the specific and the global" (Hall and Chen 1996: 407).

The issue of agency is arguably more relevant in today's 'global disorder' – "*no one seems now to be in control.* Worse still – it is not clear what 'being in control' could, under the circumstances, be like" (Bauman 1998: 58) – than in the apparently 'ordered' period of the Cold War which for much of the time coincided with the post-war economic boom and Fordist modes of accumulation in the West. The increasing, though geographically uneven, diffusion of more flexible modes of accumulation as part of and alongside other processes of globalisation, together with a new round of time-space compression, have led to a growing concentration (the degree of which is a hotly debated issue) not only of economic but also of political power in the hands of transnational corporations and global elites that are not bound by the limits of time and space in the same way and to the same degree as in the past and not trapped in the power politics of the two Cold War blocks. The contemporaneous reorganisation of spatial and temporal relations has led to new problems of anchoring social relations, social formations and political agency. Bauman points out that there is

> a new asymmetry emerging between the exterritorial nature of power and the continuing territoriality of the 'whole life'. [...] Thanks to new 'bodylessness' of

> power in its mainly financial form, the power-holders become truly exterritorial even if, bodily, they happen to stay 'in place' (ibid.: 9, 19)

These processes have been accompanied by a creeping erosion of ideals *and* realities of participatory democracy. National standards, regulations, protective structures and entitlements which had been created through centuries of democratic struggle (and not as a by-product of the development of the 'free' market) are coming increasingly under pressure in the name of the alleged 'necessities' of globalised markets and supply-side economics.

The preliminary analysis of the problematic of new distributive patterns of political agency below considers four different dimensions: the post-sovereign character of nation-states, the apparent lack of political alternatives within nation-states, the question of strategies of local mobilisation and the crisis of global governance.

Post-sovereign nation-states

The nation-state – which had been a less than perfect framework for the protection of the interests of large sections of its populace and as a host to democracy in most countries already *before* the shift to more flexible modes of accumulation – has come under increasing pressure to preserve its sovereignty in many respects. That national sovereignty has been curtailed is widely accepted but there seems to be much less agreement about the impact of specific changes and forces on the nation-state, the degree to which sovereignty has been undermined and concerning the uneven impact on different states. Giddens, in agreement with most observers, assumes that "[e]conomic influences are certainly among the driving forces – especially the global financial system" that erode the sovereignty of the nation-state (Giddens 1999: 14). But Giddens, and this is critical for the assessment of possibilities, conditions and limits of agency, also insists that these forces "aren't like forces of nature. They have been shaped by technology, and cultural diffusion, as well as by the decisions of governments to liberalise and deregulate their national economies" (ibid.). Many of the changes which have taken place over the last few decades and which are unspecifically attributed to 'globalisation' are *not* the outcome of the free play of market forces – neither in its neo-liberal versions nor in the traditional version of the 'invisible hand' of Adam Smith. Gurtov (1999) discusses, for example, the formation of the 'Trilateral Commission' in 1972, one of the best-known and most-influential 'Corporate-Globalist organizations'. Apart from the initiator, David Rockefeller, chair of the board of Chase Manhattan Bank, the commission included representatives from different walks of life (business, media, politics and labour) from the United States, Western Europe and Japan, among them several members of the Carter administration – most notably Carter himself (Gurtov 1999: 32-3; for details see Sklar 1980). The objective of the Trilateralists was "to promote intercapitalist cooperation through the removal of barriers to trade and investment among the three major regions represented in the organization and in the socialist and underdeveloped countries as well" (Gurtov 1999: 33).

Interdependence on a global scale has clearly limited the ability of nation-states to conduct their affairs without any reference to other players that operate above state level, including international institutions and MNEs/TNCs. But any superficial historical analysis would show that during the era of the modern nation-state for many of these sovereignty has always been severely limited: formerly colonised countries have only relatively recently achieved nominal political independence, the countries of the Soviet bloc were far from sovereign in their decisions (the same could, perhaps to a lesser degree, be said about the Western block) and small and weak states have always been influenced if not downright dominated by powerful states. Many of the currently voiced concerns over the precarious future of the nation-state seem to be based rather on some ideal-type notion of the nation-state than historical evidence.

The challenges to the sovereignty of nation-states which undoubtedly exist should not lead to the assumption that nation-states are helpless victims of the onslaught of exogenous forces – interestingly, politicians do not very often actively oppose the view that they are rather helpless in the face of those global economic forces.). This would clearly amount to an underestimation of both the functions and powers states continue to have. Hobsbawm accepts that states are now weaker than before but reminds us – and again this is crucial for the issue of agency – that it is necessary to distinguish

> between what states cannot do, on the one hand, and what they could do if they wanted to. Much of what governments refrain from doing is rejected not because it is ineffective – for instance, economic protectionism and a degree of self-sufficiency can work – but because, for various reasons, governments do not desire it. (Hobsbawm 1996: 62)

National governments might not 'desire' to do *certain* things but the continued ability of action of (strong) states cannot be denied. For those who would like to turn to the state in response to threats of economic globalisation, there is little hope, however, because

> [m]ore than the 'enemies' of economic globalization [...] powerful states have been its best friends [...] partly because of the capture of governments by advocates of an ideology of economic globalization [and] also as a result of territorial competition between states for the increasingly footloose capital they helped liberate in the first place. (Agnew 2001: n.a.)

Agnew's observation, however, must not be interpreted as a plea for weak states.

Lack of political alternatives within nation-states

The perceived (or real) meaninglessness of national elections and the declining degree of party-political participation in Western democracies reflect the widespread feeling that the mainstream political parties do not offer real alternatives anymore – at least not when in power (Hutton 1996: 17) – or can make no substantial difference anyway. The partial loss of national sovereignty and autonomy to the agents of a seemingly faceless and placeless global economy and politicians who present globalisation as a quasi-natural force have only served to intensify this feeling. The underlying problems are of a more general nature, however, and cannot be reduced to extranational forces or individual politicians. It is, therefore, similarly important to consider what Hobsbawm characterises as 'infranational forces' which have attenuated the links between citizens and public affairs, at least in democratic states. Among the changes listed by Hobsbawm are, for example, the decline in ideological mass parties, politically mobilising electoral 'machines' and other organisations for mass civic activity such as labour unions which has coincided with the spread of consumer individualism and the respective values. These changes do not only have serious implications for the role the state can assume but also for the hegemonic project of the transnational historical bloc referred to above:

> This withdrawal of citizens affects the legitimacy of national governments and their functions, as well as the demands which they can make upon the country's inhabitants. [...] The state is weakened when it is not identified with a common good, or when only individual advantage and not common interest are recognized [...]. The currently widespread dislike of 'bureaucracy', 'government', 'excessive state interference' and the like may contain both a justified critique of the uses and modes of state power and the unjustified assumption that any limitation by superior authority of the individual's right to do what she or he wishes, is unacceptable. We need to distinguish clearly between them. (Hobsbawm 1996: 61)

The situation described by Hobsbawm is very likely to foster – consciously or unconsciously – popular consent to anti-statist and anti-collective neo-liberal agendas which are pursued by the hegemonic bloc which has no particular interest in strong states and 'big governments'. Adam Smith, often alluded to by neo-liberals, was fully aware that 'nations' cannot be left exclusively to the free play of market forces. A fact that is less often referred to than the working of the "invisible hand" (Smith 1952: 194) is Smith's equally strong conviction that one of the duties of the "sovereign or commonwealth", i.e. the state,

> is that of erecting and maintaining those public institutions and those public works, which, though they may be in the highest degree advantageous to a great society, are, however, of such a nature that the profit could never repay the expense to any individual or small number of individuals, and which it therefore cannot be expected that any individual or small number of individuals should erect or maintain. (ibid.: 315)

The specific balance between public and private, state and civil society, government and market depends, of course, crucially on the situation in a particular country. Therefore, the task in any given national context would be to identify and support those forces that aim for a balance that is in the interest of a majority of the population.

Strategies of local mobilisation

In view of the global reach of increasingly mobile capital, multinationally, transnationally and in some case truly globally operating corporations and the apparent lack of willingness and/or the inability of national governments to challenge the excesses of the neo-liberal project, communities might consider to turn to local strategies both in their search for protection against the onslaught of globalisation and in their attempt to exploit relative locational advantages over other competitors to attract investments and to create jobs. The spectrum of alternative measures and the degree of choice offered within the global market are, however, severely circumscribed. Companies and countries do have some room of manoeuvre, they "can take the edge off certain effects or give additional edge to others; but one thing they cannot do is opt out of the global marketplace" (Dahrendorf 1996: 27).

Under conditions of flexible accumulation, time-space compression and exacerbated competition, national and local communities, large and small companies, and individuals in the developed and the developing world have all been forced to develop new strategies to prosper – and in many cases to survive. The intrinsic problem of any *local* strategy is that it would have to be responsive to intensified global competition – corporations can and do choose *globally* from thousands of *local* communities competing for investment and jobs, and equally being 'devoted' to supply-side economics. Regardless of all the fashionable talk of flexibility and the need for mobility the situation is – more often than not – one of *localised* labour vs. *placeless*, *locally unbound* capital and its managers. Time-space compression, as already argued above, does affect everybody but in very different ways. On the one hand, you have locally recruited employees "burdened as they might be by family duties, home ownership and the like – [who] could not easily follow the company once it moves elsewhere" (Bauman 1998: 8). On the other hand, you have the shareholders in the interests of which the companies are run:

> In principle there is nothing space-determined in the dispersion of the shareholders. They are the sole factor genuinely free from spatial determination. And it is to them, and to them only, that the company 'belongs'. [...]. The company is free to move; but the consequences of the move are bound to stay. Whoever is free to run away from the locality, is free to run away from the consequences. These are the most important spoils of victorious space war. (ibid.: 8-9)

Local actors, in other words, have to face an only seemingly paradoxical situation: "Spatial barriers have collapsed so that the world is now a single field within which capitalism can operate and capital flows become more and more sensitive to the relative advantages of particular spatial locations" (Waters 1995: 57-8). This situation produces a permanent pressure on local communities to provide better conditions for prospective investors. Local problems, therefore, cannot be solved locally on a permanent basis and would need to be re-integrated into a global macro-economic system. Any realistic approach to local problems would therefore have to reach for the translocal and transnational level at which the global economy operates. This, however, is a problem to which no satisfactory answer has as yet been provided. The intermediary, national level which would have to work as a hinge between the local and the transnational has come, as outlined above, under increasing pressure from the forces of global capital (and disaffected citizens) itself and the remaining opportunities, circumscribed as they may be, are often not realised by national governments for various reasons.

Crisis of global governance

The existing institutions and structures of the nation-state, on the one hand, cannot simply be scaled up in the hope to regulate increasingly worldwide processes and transactions along the lines of well-tried national policies. This is particularly problematic because national identities continue to be a very important source of extra-local political mobilisation. On the other hand, existing institutions with a global reach like the United Nations, the International Monetary Fund, the World Bank, the World Trade Organisation continue to be dominated by a small number of the most powerful states and the business interests they represent – they can therefore not be expected to work in the interest of some global public good. Woods describes the differences between strong and weak states: "Strong states are trying to shape institutions to manage financial crises, powerful NGOs, and globalizing firms. Weak states are trying to survive increasingly precarious and changeable economic circumstance" (Woods 2001: 297). Giddens seems to put some hope in the European Union and similar organisations that might emerge in the future when he discusses the possibilities to foster democracy above the level of the nation-state because "it is pioneering a form of transnational governance. [...] The countries which have entered the EU have voluntarily given up some of their sovereignty in order to do so" (Giddens 1999: 80). The specific form of transnational governance embodied in the EU does not, however, necessarily mean that the power of global capital could be – or is – controlled more effectively if the need arises, than by a particular nation because the EU has been captured by advocates of an ideology of economic globalisation perhaps more strongly than some of its member states. Moreover, the previous history of the EU has shown that it is rather difficult to make any substantial impact on decision-making structures and institutions that are far removed from national electorates.

Woods proposes an interesting argument with reference to the degree of faith most states seem to invest in international institutions in their common search for greater stability and predictability when he writes that

> in recent years virtually every state in the world has joined at least one regional trade grouping [which] underscores the search for new ways to manage globalization. At the same time, regionalism highlights the scepticism of many states about international institutions, and their fears that institutions are too dominated by powerful states, or unlikely to constrain them. (Woods 2001: 297)

Conclusion

Concluding, it could be argued that the global economic system and the flexible mode of accumulation on which it currently rests tend to disempower, though very unevenly, both the national and the local, and to undermine increasingly what there had previously been in terms of political agency and democratic accountability. What seems to be needed analytically in view of progressive globalisation is an increased awareness of local particularities to avoid over-generalisations, simultaneously an acceptance of the global context in which the relations between local place and global space are being reconfigured and a sharper historical awareness of the continuities between past and present. Territorial or place-based patterns will not lose their relevance for the distribution of political agency despite the high degree of time-space compression. But, at the same time, it cannot be denied that capital "continues to dominate, and it does so in part through superior command over space and time, even when opposition movements gain control over a particular place for a time" (Harvey 2001: 238-9). Or, as Bauman puts it:

> The *global* dimension of the investors' choices, when set against the strictly *local* limits of the 'labour supplier' choice, provides that asymmetry which in its turn underlies the domination of the first over the second. (Bauman 1998: 105)

Works Cited

Agnew, John (2001). "The New Global Economy: Time-Space Compression, Geopolitics, and Global Uneven Development," *Working Paper* 3, 2001 (Lecture presented at the Center for Globalization and Policy Research, UCLA). <www.sppsr.ucla.edu/cgpr/docs/workingpaper3.doc> (23 February 2005).

Bauman, Zygmunt (1998). *Globalization: The Human Consequences*. Cambridge: Polity.

Bruff, Ian (2005). "Making Sense of the Globalisation Debate". *The British Journal of Politics and International Relations* 7.2, 261-280.

Dahrendorf, Ralf (1996). "Economic Opportunity, Civil Society and Political Liberty". *Social Futures, Global Visions*. Ed. Cynthia Hewitt de Alcántara. Oxford & Cambridge, M.A.: Blackwell & UNRISD, 19-38.

Giddens, Anthony (1999). *Runaway World: How Globalisation Is Reshaping Our Lives*. London: Profile Books.

Gurtov, Mel (1999). *Global Politics in the Human Interest*. Boulder et al.: Lynne Rienner.

Hall, Stuart & Kuan-Hsing Chen (1996). "Cultural Studies and the Politics of Internationalization: An Interview with Stuart Hall by Kuan-Hsing Chen". *Stuart Hall: Critical Dialogues in Cultural Studies*. Ed. David Morley & Kuan-Hsing Chen. London & New York: Routledge, 392-408.

Hall, Stuart (1991). "The Local and the Global: Globalization and Ethnicity". *Culture, Globalization and the World-System*. Ed. Anthony King. Houndmills & London: Macmillan, 19-39.

Hall, Stuart (1996a). "The Problem of Ideology: Marxism without Guarantees". *Stuart Hall: Critical Dialogues in Cultural Studies*. Ed. David Morley & Kuan-Hsing Chen. London & New York: Routledge, 25-46.

Hall, Stuart (1996b). "The Meaning of New Times". *Stuart Hall: Critical Dialogues in Cultural Studies*. Ed. David Morley & Kuan-Hsing Chen. London & New York: Routledge, 223-37.

Harvey, David (2001 [1990]). *The Condition of Postmodernity: An Enquiry into the Origins of Cultural Change*. Cambridge & Oxford: Blackwell.

Hay, Colin and Ben Rosamond (2002). "Globalization, European Integration and the Discursive Construction of Economic Imperatives". *Journal of European Public Policy* 9.2, 147-167.

Hay, Colin and David Marsh (2000). "Introduction: Demystifying Globalization". *Demystifying Globalization*. Ed. Colin Hay & David Marsh. Basingstoke: Macmillan, 1-17.

Hobsbawm, Eric J. (1996). "The Future of the State". *Social Futures, Global Visions*. Ed. Cynthia Hewitt de Alcántara. Oxford & Cambridge: Blackwell & UNRISD, 55-66.

Hutton, Will (1996). *The State We're in*. London: Vintage.

Marx, Karl (1981). *Grundrisse: Foundations of the Critique of Political Economy (Rough Draft)* (Translated with a Foreword by Martin Nicolaus). Harmondsworth: Penguin.

Sklar, Holly K. (ed.) (1980). *Trilateralism: The Trilateral Commission and Elite Planning for World Management*. Boston: South End Press.

Smith, Adam (1952 [1776]). *An Inquiry into the Nature and Causes of the Wealth of Nations*. Chicago et al: Encyclopaedia Britannica.

Waters, Malcolm (1995). *Globalization*. London & New York: Routledge.

Woods, Ngaire (2001). "International Political Economy in an Age of Globalization". *The Globalization of World Politics*. Ed. John Baylis & Steve Smith. Oxford: OUP, 278-98.

Tarja-Liisa Laaksonen

Situating Liberal "Organizational Development" and "Good Governance" Discourses in Globalisation and Political Contexts

Introduction

This paper engages with current discourses of "organisational development" and "good governance" as they are expressed in organisation studies and social sciences, and calls into question some of their major themes.

Worrying in these discourses is that they assume that clearly structural and political problems could be resolved by addressing them at individual, organizational, or governance levels. Reflecting on the current governance and organisation discourses, and for the analytical purposes of this paper, I distinguish certain clusters of analysis. I argue that the current governance and organisation discourses are inevitably situated in globalisation and political contexts, and those are further set in a complex interplay of local and cultural considerations, as well as aspects of domination and human agency. Political and globalisation perspectives are, however, underlying these other clusters. They, in turn, indicate serious democratic deficits, and show that governance and organisational development issues are always critically political.

Politcs of Globalisation and Governmentality

Globalisation is a term often used when referring to the global reach of the economy and its underlying regimes of thought. The fact that the term is referred to in such a way, reflects and reproduces the universalising tendencies of the dominant neoliberal rationality of governments. The global economic system is now based on the neoliberal ideas of non-government of society, and the promotion of freedom and cultural reforms (Dean 1999: 150). This mode of thought extends to all spheres of life through global interconnectedness, which is also referred to as globalisation. Thus, due to its very characteristics, the neoliberal regime of thought is not only economic and national, but also social, political, cultural and global. This dominant underlying regime of thought emphasises the acknowledgement of risk and uncertainty as critical to liberal freedom and enterprise.

Comprehension of the notion the "risk society" as sustained by Beck (Beck 1992) is considered as limited to de-traditionalisation and modernisation discussions, because it does not sufficiently take into account the political governmentality implications of the risk as a discourse. O'Malley (O'Malley 2000: 480) distinguishes between risk as something, which can be calculated and to a certain extent managed. Risk is a statistically and professionally estimated field of expert scientific governance. Uncertainty is a mode of self-government required of the liberal, rational, free and calculating subject. Individuals carry and resolve the uncertainty through enterprising. Thus, uncertainty requires individuals as subjects who govern themselves as rational and responsible agents. "Uncertainty is central to profitability and good economic governance" (O'Malley 2000: 463) and has a role as a "discourse" in neoliberal thought. The neoliberal pillar for managing uncertainty is the cult of enterprising, which is also inserted in all spheres of life. It highlights the role of enterprise culture, active subjects, and innovating entrepreneurers. An entrepreneur has the responsibility and skills to manage uncertainty.

Drawing from this new era of neoliberal rationality – or 'advanced liberal government' and 'reflexive government' as its variations are named by Dean – has been argued as signifying "death of the social" (Rose 1996; Dean 1999: 151). These regimes of thought do not intend to govern through society anymore by implementing traditional mechanisms, techniques of agents of government. The difference is that they seek to govern through the society and transform it through uncertainty and fear (Dean 1999: 196-197).

Clearly, discussions about governmentality and neoliberal globalisation provide a basis for understanding the discursive constructions of the "organisation and governance" phenomenon, in the way as I approach the phenomenon in this paper. "Governmentality" examines the "mentalities, arts, and regimes of government and administration that have emerged since early modern Europe" (Dean 1999: 2). It is a way of analyzing the political economy of liberalism as a rationality of thought. Specifically, Foucault's post-structuralist analysis shows how neoliberalism is a historically formed discourse, and his governmentality analysis is a critique of liberalism (Olssen 2003: 189). Governmentality is defined as the "conduct of conduct" or "any attempt to shape with some degree of deliberation aspects of our behaviour according to particular sets of norms and for a variety of ends" (Dean 1999: 10). This forms here the basis by means of which the analysis of the management culture literature and critical organisational culture literature emerges and can be grasped. Due to a lack of wider analysis about the underlying regimes of thought, these academic fields are deficient of political dimensions, which the globalization and governmentality academic discussion explicitly contains (Miller and Rose 1990; Burchell, Gordon et al. 1991; Rose 1991; Rose and Miller 1992; Rose 1993; Rose 1994; Barry, Osborne et al. 1996; Rose 1996; Barnett 1999; Dean 1999; Rose 1999; O'Malley 2000; Rose 2000; Lemke 2001; Peters 2001)

I hold that the "neoliberal globalisation" discourse is the dominant regime of thought underlying the present-day discourses of "organisation and governance". My intention is, however, to explore how culture and individual agency and the "collective possibility of transforming society and establishing new forms of social controls on the economy" exist within such a context (Touraine 2001: 6). Antagonisms are found against the neoliberal discourse and its derivatives of entrepreneurialism, uncertainty, and individualism discourses.

I intend to explore an extension of the term "globalisation" so that it does not, by definition, automatically cover the imposition of hegemonic economic to social, political and cultural. Furthermore, I question if neoliberal society or "globalisation" in creation is purely ideological, as proposed by Touraine (2001), and thus cannot convey further enduring and overwhelming manipulations and transformations in social, political, and cultural domains. Market-driven, individualistic and utilitarian neoliberal discourses create obvious conflicts, which engender struggle and strengthen agency against the adoption of global discourses. Moreover, Laclau and Mouffe (Marsh and Stoker 1995: 119-123) suggest that "discourses do not simply mirror processes occurring in economy. Rather, discourses incorporate elements and practices from all parts of society".

They refer to this combining of various different elements as "articulation". These social antagonisms create "political frontiers" and "political subjectivity". Laclau and Mouffe argue that 'no matter how successful a particular political project's discourse might be in dominating a discursive field, it can never in principle completely articulate all elements, as there will always be forces against which it is defined (Howarth 2000: 103).

Drawing on the multiple and contradictory nature of discourses, globalisation can be understood as composed of diverse political and economic formations - including multiple capitalisms and transformations. Globalisation is then understood as complex entanglements of political economy, rationalities, and cultural politics. By this definition, globalisation then inevitably encompasses mutually informing relationships of regimes of thought, technologies of domination, and resistance-agency in culturally specific conditions. By the same token, we could speak about interconnected regimes of rationality, technologies of ruling, and technologies of self. For the purposes of this paper this definition covers the dynamics and interplay between global, local and subjective forces in the reproduction and maintenance of the "discourse of organisation and governance" as a phenomenon of governmentality in the context of globalisation.

In sum, in political and global spheres, neoliberalism and advanced liberalism are the present-day underlying rationalities of thought for organization and governance development discussions. In neoliberalism, the state has gained a connotation as an active agent in creating the best market conditions and by creating laws and institutions for its maximum operation. Neoliberalism also intervenes in the creation of an individual by enhancing the enterprising and competitive model. The individual is then manipulable, and the state plays an important role in the shaping of this entrepreneurial figure. Thus, neoliberalism is governmentality of "freedom, choice, consumer sovereignty, competition, individual initiative, compliance and obedience" in which the state again has an important role as a regulator and in conditions where all social relations are regarded as relations of exchange (Olssen 2003: 199). Notably, neo-liberalism introduced the recovery of the market in all spheres of life, but it also claims that change is not to be rational but must be carried out according to cultural values, rules and norms – which are "best condensed into the cultural form of enterprise and consumer" (Dean 1999: 164). Thus, we see now that neoliberalism or its more recent mutation "advanced liberalism" governs through experts and through aspirations of freedom that citizens have. It also utilizes moral and community loyalty in the administration of conduct, and operates from the distance between the political representation and other social actors (Rose 1996: 41, 54).

I argue that the present-day organization and governance discourses are too silent about the underlying neo- and advanced liberal political and globalization processes, which are the principal structures for the so called "organizational development" and "good governance" work. Consequently, I call for a more serious inclusion of structural dimensions into discussion and analysis.

Local and cultural

Often related studies and the social sciences assume that the concepts "organisational development/culture" and "good governance" are universally understood and have a shared meaning worldwide. These discourses are clearly informed by the Anglo- and Eurocentric management and social sciences literatures. They often treat difference and culture as something exotic. Contradicting this, they however assume that organisations or enterprises can be treated with the same management and governance ideas all over the world. Thus, they usually lack comprehensive local and cultural perspectives. The homogenous universality and unproblematic globalisation of the neoliberal or advanced liberal regime of governance thought is assumed.

However, I argue that culture is also always governmental since it implies power and practices, and the ordering of people and things. Thus, also globalization also needs to be seen in complex relations between power, culture and different spatialities. Then, cultural importation and imposition cannot be generalized. Instead, we can observe that the production of something quite new occurs when cultural forms 'traveling across borders, often become comprehensively and unpredictably reinterpreted and re-customized to serve very particular local purposes' (Yurchak 2000: 412). Furthermore, individuals are not only national subjects, but have a multiplicity of subjectivities in various sites and scales and thus "there is no longer a general subject of culture" but the globalization process "reterritorializes the normative individual subject" (Barnett 1999: 390). In the advanced liberalism ethos individuals are increasingly addressed as "members of communities". In these respects culture and governmentality need to be seen in multiple spatialities and temporalities, in which organizations and spaces are one set of means of exercising power.

However, we need to be careful in considering certain lived spaces as purely global or local. Rather, it is important to look at the globalization as a more nuanced configuration of various types of spaces and their intersections. There is no singular, homogenous, and abstract space of globalization. A "world culture or global culture" is something we can not even imagine – it is an abstract impossibility. Instead, globalization is a process. In the globalization process, lives may be experienced as out of place, when the localized affective identities of nationhood, ethnicity, and religion contrast with the appealing cosmopolitan globalism (Crang 1999: 171). Similarly, local cultures cannot be imagined as "uncontaminated or self-contained" or not affected by other cultures. (Bhabha 1998: 54).

"Culture" examined in terms of governmentality thus receives new dimensions. Traditionally, culture has been defined "as a set of signifying practices" where it became both an object and an instrument of governmental policy. However, once we begin to see "culture" in relation to governmentality as an integral part of the policing process we can extend the analysis towards power and culture. Then we can, for example, examine how neo-liberal or globalization strategies operate in a specific cultural context, or how culture functions as the intersection of policy and ethos – that is "the practices of the self on the self and the technologies of subjectification". Thus, the objective of cultural studies shifts from language bound general mechanisms and signifying practices of a certain culture towards technical, discursive, truth-telling governmentality practices found in a specific cultural setting and historical moment. (Bratich, Packer et al. 2003: 5-6). Then, it is easier to see how today cultures rather mingle and interact with each other. It is a global and local fusion. Suyngedouw argues that the spatial scale is rather a produced process, which is continuously heterogeneous, conflicted, and contested. This scale is discursively and materially a place where power relations are contested, negotiated and regulated (Suyngedouw 1997: 140).

Bhabha's concept of "hybridity" approaches the same understanding when he describes the process of "construction of cultural authority within conditions of political antagonism and inequality (...) and where hybrid strategy or discourse opens up space of negotiation". The emerging negotiation is not assimilation or collaboration, but instead hybrid agency is born. (Bhabha 1998: 58). Here, Appadurai's diagram of various spheres or scapes is useful. We can find ethno-scape, media-scape, techno-scape, finance-scape, and idio-scape. These spaces are not limited areas, but instead like landscapes, which are formed of different dimensions or scapes which become realized in an experienced place (Appadurai 1996). In these various networks of spatial scales and scapes, economics, politics, social issues, and identity formation become shaped (Appadurai 1990). Thus a concrete moment of lived place can be experienced as a site where "globalization comes together and also falls apart" and Appadurai's different "scapes" offer an imagined space in which the local is embedded but of which it is a "unique configuration". Thus, globalization becomes to be seen as performed when people negotiate a variety of spaces. (Crang 1999: 175).

In sum, I argue that in our globalized world "culture" is becoming a hybrid and fluid mixture of global and local multiplicities and spaces or scales. We just need to discover and recognize those specific hybrids of discourses, spaces of culture, and scapes in any organization or governance setting. Thus, it seems not viable that certain globally homogenized "organization or governance" ideas could be even imposed on.

Domination versus resistance-agency

The "organisation and governance discourses" also assume that even if workers or organisations are analysed with regard to their ways of resistance, they are in somewhat powerless positions. Most of these positions identify 'organisation and governance discourses' as a tool or a management technology involved in the creation of appropriate and "excellent" cultures, including notions of the self-regulating and entrepreneurial self in various institutional environments. In such contexts, these discourses centre on fixed subjectivities, control over organisation and organisational subjects, as well as focus on the domination of entrepreneurial discourses over subjectivities.

The liberal concept of "freedom" is in a key position in shaping the regimes of thought or political rationalities and ultimately management discourses, and the ways in which they in turn outline the conduct of conduct. In advanced liberalism freedom is the formula of rule. Through discourses of freedom, self-disciplined citizens are produced. They are responsible agents working eagerly for their own development. This advanced liberal governmentality does not, however, function via direct domination and imposition, but through hope and promise. Those are the promises of incorporation and inclusion, which are supposedly achieved via worker commitment, self-government and agency. This advanced liberalism usage of "agency" for its objectives is controversial and needs some clarification.

In the liberal discourse, duty of the self is to function as a moral agent and as a rational choice and economic actor. In these freedom circuits of the self, the risk management is also then transferred back to individuals (Peters 2001: 61). Similarly, in organizational life, success of employment or career development and risk of failure or unemployment are transferred back to individuals. The new valorization of the self-actualized subject coincides with the emergence of the neo- and advanced liberal redeployment of the "free subject" as an instrument of governmentality, and it coincides also with the notion of freedom and free conduct as the principles by which the government is to be rationalized and reformed (Dean 1999: 155).

Then, these technologies of agency seek to improve and enhance our possibilities for participation, agreement and action. They emphasize the "voice" and "representation" of user groups or partners to negotiate over their needs. The objective of this agency work is to enable the 'targeted populations' or the "excluded" ones to take control of their own risks and act for their development. (Dean 1999: 167-168). It thus manages the themes of freedom and agency in a reflexive way in its deployment of surveillance and regulation of the agency through which it functions (Dean 1999: 195).

Cruikshank's examples illuminate these discourses and related agency technologies. She distinguishes phases of "cultures of self" discourses: In the 1970's the theme of the "self-governance technologies" discourse was that of "positive thinking", in the 1980's it was that of "empowerment", and in the 1990's "enterprising" (Cruikshank 1996: 247). She argues that towards the turn of the century we were experiencing the discourse of "self-esteem". Today, we live the "ethico-community" discourse as illustrated by Rose (Rose 2000a). Or rather, we often live in combinations of these discursive technologies. Most importantly, however, it is crucial to analyze how the lack of power and control in the personal life and in the world is increasingly attributed to a lack of these themes of positive thinking, empowerment, enterprising, self-esteem, and ethico-community responsible conduct. Instead of addressing the structural issues, the political liberal environment continues to insert these new themes which account for the individuals – and their agency or lack of it –, for the ills of the society and the world.

Importantly, nowadays it is the combination of normalizing discourses, disciplinary effects and the constitution of calculating enterprising subjects that are in play (Martin 2002: 4). We are now witnessing the new concept of freedom that "obliges us to self-realization through choice and to the ethico-politics of community": enterprise, organization, or governance (Rose 1999: 273).

This "governmentality" agency is different to the agency of resistance over attempts of subjection. I argue that even if individuals continuously receive and listen to the neo- and advanced liberal governance, management and enterprise discourses, they also negotiate these dominant discourses. As a result of negotiation, the meaning of dominant discourses is only partially fixed and thus discourses are only partially fixing of subjectivities. Rather, in this flow of discourses, the construction of multiple subjectivities and selves is also in continuous movement between, for example, acceptance, adoption, collaboration, conflict, resistance and compromise. Thus, the dimensions between compliance and resistance can be parts of wider social play.

Conclusion

This discussion provides a frame for an "organisation and governance development" research strategy: It is necessary to look at how the politics of neo- and advanced liberal discourses and globalization discourses introduce redefinitions of "organisation and governance development", culture, subjectivity, and conduct.

Furthermore, they impose management and enterprise culture assimilation, and extend market language to all possible activities. However, such global constructions of organization and governance cultures are neither possible nor viable.

Additionally, individuals and organizations "agentially play with the associated discursive practices and forms" so that "resulting outcomes are complex and varied" (Doolin 2002: 385-387). So, globalization and (organizational and governance) cultural imposition are not capable of simply introducing a new system of signification or order of people and things – depending on how we want to define culture – into a physical public organisation or enterprise space. Instead, a redefinition of that space, personal subjectivites, and power relations results. Thus, resistance and agency are at play when the intentions of particular, new versions of social realities and the construction of subjectivities are developed.

When these aspects are taken into account, the "organization and governance development discourses", establishment of fixed values, creation of committed but self-regulating organisational subjectivities, and binaries that divide human life in work, organisation and other life, are broken down. Human beings are not one-dimensional objects. They live through shifting multiple subjectivities in constantly changing organisation and governance situations. Their political subjectivity is well alive and neoliberal globalisation resisted. These observations indicate global and political implications of the whole "organisation and governance development" discourses. When these complexities are considered, the whole liberal concept of the present-day management, organization, and governance development discourses – in the way they appear and are promoted – is contested. „'‘

Works Cited

Appadurai, Arjun (1990). "Disjuncture and Difference in the Global Cultural Economy". *Theory, Culture & Society* 7, 295-310.

Appadurai, Arjun (1996). *Modernity at Large: Cultural Dimensions of Globalization*. Minneapolis, Minn., University of Minnesota Press.

Barnett, Clive (1999). "Culture, Government and Spatiality: Reassessing the 'Foucault Effect' in Cultural-Policy Studies". *International Journal of Cultural Studies 2*, 369-397.

Beck, Ulrich (1992). *Risk Society: Towards a New Modernity*. London, Sage.

Bhabha, Homi K. (1998). "Culture's in-between". *Questions of Cultural Identity*. Ed. Stuart Hall & Paul du Gay. London: Sage, 52-60.

Bratich, Jack Zeljko & Jeremy Packer et al. (2003). "Governing the Present". *Foucault, Cultural Studies, and Governmentality*. Ed. J. Z. Bratich, J. Packer & C. McCarthy. New York, State University of New York Press, 3-21.

Crang, Mike (1999). "Globalization as Conceived, Perceived and Lived Spaces". *Theory, Culture & Society* 16, 167-177.

Cruikshank, Barbara (1996). "Revolutions within: self-government and self-esteem". *Foucault and political reason. Liberalism, neo-liberalism and rationalities of government.* Andrew Barry, Thomas Osborne & Nikolas Rose. London: Routledge UCL Press.

Dean, Mitchell M. (1999). *Governmentality: Power and Rule in Modern Society*. London. Thousand Oaks: Sage.

Doolin, Bill (2002). "Enterprise Discourse, Professional Identity and the Organizational Control of Hospital Clinicians". *Organization Studies*. 23, 369-390.

Foucault, Michel (1991). "Governmentality". *The Foucault Effect: Studies in Governmentality, with two Lectures by and an Interview with Michel Foucault*. Graham Burchell, Coin Gordon, Peter Miller & Michel Foucault. London: Harvester Wheatsheaf.

Martin, Peter (2002). *The Constitution of Power in Landcare: a Post-Structuralsit Perspective with Modernist Undertones*. Centre for Rural Social Research, Wagga Wagga: Charles Sturt University.

Olssen, Mark (2003). "Structuralism, Post-Structuralism, Neo-Liberalism: Assessing Foucault's Legacy". *Journal of Education Policy* 18.2, 189-202.

Peters, Michael (2001). "Education, Enterprise Culture and the Entrepreneurial Self: A Foucauldian Perspective". *Journal of Educational Enquiry* 2.2, 58-71.

Rose, Nikolas (1996). "Governing 'Advanced' Liberal Democracies". *Foucault and Political Reason. Liberalism, Neo-Liberalism and Rationalities of Government*. Andrew Barry, Thomas Osborne & Nikolas Rose. London: Routledge. UCL Press, 37-65.

Rose, Nikolas (1999). *Powers of Freedom: Reframing Political Thought*. New York, Cambridge: Cambridge UP

Rose, Nikolas (2000). "Community, Citizenship, and the Third Way". *American Behavioral Scientist* 43.9, 1395-1411.

Suyngedouw, Erik (1997). "Neither Global nor Local: 'Glocalization' and the Politics of Scale". *Spaces of Globalization: Reasserting the Power of the Local. Perspectives on Economic Change*. Kevin R. Cox. New York: Guilford Press, 137-166.

Yurchak, Alexei (2000). "Privatize Your Name: Symbolic Work in a Post-Soviet Linguistic Market". *Journal of Sociolinguistics* 4.3, 406-434.

Jörg Meyer

Agency Lost

Scientific Respectability and Real World Relevance as Pitfalls of Contemporary International Thought

Introduction

Although there is a lot of talk about actors and actions, the academic discipline of International Relations (IR) largely disregards agency. This is not to say that one cannot find the term in books and journals. Nor do I want to deny that there are some remarkable efforts to introduce a concept of agency into the study of international politics (cf. Debrix 2003). But my claim is that the prevailing ways in which the world is portrayed in IR give the reader hardly an impression that human actions, let alone their deeds and choices, make a difference.

If this is actually the case then obviously not as a result of a commitment to some anti- or a-human postmodernism and the catchphrase of the "death of the subject". On the contrary, theories of IR often proceed from the presumption of "sovereign" actors, thus affirming the modern notion of the autonomous and rational subject. Moreover, in so far as research is designed not only to discover causal mechanisms and/or to predict outcomes, but also to provide instructions for problem-solving, scholars of IR explicitly suppose policy makers with the will to effectuate, for instance, security or economic prosperity.

The aspiration for scientifically respectable as well as relevant knowledge, however, forecloses the possibility to tell stories about international politics which give way to agency. What becomes disseminated is a bulk of "testable" hypotheses, (more or less directly) designated for the management of problems in the "real world". Yet, as Hannah Arendt pointed out, an orientation towards calculation and control is inconsistent with agency. "Predictions of the future are never anything but projections of present automatic processes and procedures, that is, of occurrences that are likely to come to pass if men do not act" (Arendt 1969: 7). This does not mean that deeds are to be understood as manifestations of pure improvisation or mere chance. But what has to be avoided in view of agency, simply spoken, is an approach which makes people, in one way or another, into automatons, or to employ an image from Jacques Derrida's (1992: 23) discussion of legal judgement, into "calculating machines".

In the following, rather than trying to outline a concept of agency, I will ask how the capacity of acting, "of beginning something anew" (Arendt 1958: 9) gets lost in contemporary stories of IR. For this purpose, I shall start with a brief survey of some "epistemological"/"methodological"[1] premises which enable the discipline both to locate itself inside a respected enterprise called science and to take a position qualified for providing seemingly neutral knowledge about world politics. While the appeal to disciplined research may sound reasonable, the search for general facts and mechanisms of international life is hardly concealed with the by now common slogan that people create and change social structures. Hence, I go into the current preoccupation with norms in IR in order to show how the routine acknowledgment of a capacity to initiate and to refuse vanishes in favour of causal explanations. This leads me, finally, to the question whether reasons are or should be understood as causes, which has become some sort of focal-point for discussing the possibilities of agency.

Epistemology/Methodology and Ontology

"Can the study of ideas in international politics be made scientifically respectable?" For Alexander Wendt (2000: 165), the question is central to what has been named the Third Debate of IR, and his answer reads yes. To scholars not being at home in IR, the question may sound somewhat strange or restrictive in so far as the reference to ideas seems to suggest that there are other 'things' in international politics which are unquestionably suited for being studied scientifically. However, if one replaces "ideas in international politics" by, for instance, "society", "the social world" or "international politics" the issue should be more familiar to the reader. What is at stake, then, is the unity of science or the autonomy of the humanities respectively.

Since Wendt's argument is quite complex I will pick out only a few remarks on epistemology/methodology and ontology. In his much quoted *Social Theory of International Politics*, he subscribes to (scientific) realism as a philosophy of science and to the propositions that "the world exists independently of human beings" and that "mature scientific theories typically refer to this world" (Wendt 1999: 47). At the same time, he supports a "constructivist" or "idealist ontology" by emphasizing several differences between "natural" and "social kinds" (68-71). In short, "social kinds are social, made of ideas instantiated in practice". For

[1] While some authors in IR explicitly distinguish between epistemology and methodology (cf. Wight 2002), others use the terms more or less synonymously. I see no need to discuss the issue in this paper.

Wendt this seems to be not only a matter of 'what there is', but also important for our self-conceptions. "Since these ideas are after all our ideas, if we forget that social kinds are social then we forget that we are their makers or authors". In other words, if we forget that the social world is one of artifice we naturalize it. "As a result, rather than experiencing social systems voluntaristically, as artefacts of our design and intent, we experience them deterministically, as if they were forces of nature pressing upon us" (374).

Nevertheless, Wendt stands up for a scientific IR. When "it comes to the epistemology of social inquiry I am a strong believer in science [...]. I am a 'positivist'" (39). Considering his commitment to a scientific realist philosophy of science, for Colin Wight (2002: 36), this is an "impossible position to hold" because scientific realist and positivist accounts of science "are diametrically opposed on some very fundamental issues". While I agree with Wight that one should not confuse scientific realism with positivism, what interests me here are the ways in which Wendt tries to reconcile his propositions about social kinds and his epistemological/methodological claims.

With respect to the latter he admits that the differences between the objects of the natural and the social scientist may call for different methods of inquiry ("we cannot interview bacteria"), but emphatically rejects the call for different epistemologies. In his view, scientists "in both domains are concerned with explaining why one thing leads to another" (Wendt 1999: 372-3). Moreover, he identifies two main camps in the Third Debate of IR: a "majority who think that science is an epistemologically privileged discourse through which we can gain a progressively truer understanding of the world", and a "large minority" who dispute this epistemic status (38).

When Wendt writes that the former have become known as "positivists", his choice is clear: the study of international politics should be made respectable by accepting certain criteria stipulated by or ascribed to positivism. That is to say, we should expect of theories of IR "that they be logically coherent, rely on public available evidence, be in some broad sense falsifiable, and be true or false based on their correspondence to the world" (Wendt 2000: 171), because the "epistemic authority of any scientific study" depends on this criteria. "If there is no such evidence or if a study is non-falsifiable, then it might still be interesting as a form of art, self-expression, or revelation, but it is not an effort to know the world through 'science'" (Wendt 1999: 373).

In Wendt's story, on the one hand, "we" are the authors or makers of social kinds. On the other hand, the world can and should decide whether a social theory is true or false. I do not want to suggest that he has carelessly produced an inconsistency. In fact, Wendt acknowledges that the effort to reconcile "what many take to be incompatible ontological and epistemological positions" injects "significant tensions" into his argument (40). But how can one handle these tensions? There are two obvious ways: either one fits epistemology to ontology or ontology to epistemology. On my reading, Wendt submits his propositions about social kinds to a concept of science dominating in IR.

One central feature of this concept is the separation between science/theory and politics/the real world. Wendt is anxious to distance himself from positions who claim that theories of international relations are constitutive aspects of world politics.[2] "Theory reflects reality, not the other way around" (47). Yet, as already mentioned, he emphasizes several differences between natural and social kinds. And he goes on to write that social scientific theories, in contrast to the theories of natural science, "have the potential to become part of their world" (76). Nevertheless he insists, for instance, on the premise that "the international system confronts the IR theorist as an objective social fact" (75). The counter-argument would be, of course, to draw attention to the involvement of IR in the discourse re-producing the international system. But Wendt brushes aside such an approach from the start. Since "states are the dominant form of subjectivity in contemporary world politics", he writes, "they should be the primary unit of analysis for thinking about the global regulation of violence" (9).

Obviously, one doesn't have to equate a dominant form of subjectivity with an objective social fact. But Wendt tries hard to make such a stance more or less inevitable. In the first place, he sticks to the significance of an independent reality by pointing out that social kinds like the international system resist an arbitrary interpretation by an individual who wants to explain them. "Those who maintain social kinds never satisfy the subject-object distinction imply that professional or lay scientists can make the world anything they want".[3] Having restored the separation between theory and reality, he continues his argument by denying individuals any power to effectively challenge institutions. "Individuals do not constitute social kinds, collectives do, and as such social kinds confront the individual as objective social facts" (75). Now, one might still hold that people change the world, for better or worse, by acting in concert. Here, Wendt finally brings up the concept of "reification"[4] in order to preserve the distinction between subject and object. "Usually social kinds confront members of the relevant collectives as seemingly natural facts" (76).

In sum, Wendt works with strict alternatives. Either the theorist can make the world anything she wants or she is confronted with an objective reality. And if I, alone, do not have the capacity to transform social kinds, then they are

---

[2] For such an account, see R.B.J. Walker (1993) and Steve Smith (1995).

[3] The fact that some theories are more successful than others is frequently used in IR to obscure post-positivism or non-foundationalism (cf. Mayer 2003: 88-89).

[4] Wendt (1999: 76) quotes Berger/Luckman (1966: 89) who mean by reification the "the apprehension of the products of human activity as if they were something else than human products – such as facts of nature, results of cosmic laws, or manifestations of divine will".

objective social facts as well. Collectives, again, constitute social kinds. But in one version of his story they are the deliberate product of "our" design and in another the result of some normal or even natural[5] situation of reification. What is left out, e.g., are the questions of who speaks for "us" or which relations of power/knowledge constrain and enable certain activities.[6] And while admitting that collectives may call social kinds into question and change them, Wendt does immediately downplay the significance of such "occurrences" for ordinary live as well as standard theorizing. "In all but society's most reflexive moments, there is a distinction between subject and object" (77). But who decides when and where people (can) engage in "self-reflection"?

After being criticized for treating the state as a given (cf. Doty 2000; Smith 2000), Wendt conceded that his *Social Theory* does reify the state and added that it does so intentionally, since, in his view, it is a "good thing" to reproduce a state-centric world. "For all their faults, states are the only democratically-accountable institutions we have today to provide security and political order" (Wendt 2000: 174). While, in my view, it is legitimate to support (or to contest) the state-project, this is another matter than asserting that states are the dominant form of subjectivity. Moreover, what is at first re-presented as an objective social fact existing independently from the theorist becomes a "contested social construction" that is potentially ("in some small ways") re-produced by a theory which reifies the state (174). One might think that Wendt does not only undermine his notion of theory as a mirror of reality, but also performs the collapse of the distinctions on which his epistemological claims rest. However, in the end, he does neither modify the restrictive view of science nor the prescriptions for research in IR. On the contrary, he reinstates his warning of doing so. "The goal should be to facilitate substantive investigations rather than discourage them, to unify our knowledge rather than Balkanize it" (180).

Science and Policy

The metaphor is well chosen in so far as Balkanization is widely associated with disintegration and violent fragmentation in world politics. With regard to science, it stands for the abandonment of any objective point of view or standard for deciding between competing truth-claims. What is at stake then is not only the scientific respectability and the privileged epistemic status of the discipline,

---

[5] "Culture meets basic human needs for sociation and ontological security. [...] it is the ability to treat culture as given that enables us to go about our business. [...] The actors who make up social systems are animals with biologically constituted capacities, needs, and dispositions" (Wendt 1999: 189).

[6] For a survey of such questions see James Tully (2002). In many respects, as for instance in the emphasis on "brute facts", Wendt's approach resembles John Searle's *The Construction of Social Reality*. For a critical comment on Searle see Bernhard Waldenfels (1998).

but also the foundation of IR as an authority which helps to bring rationality and progress to foreign policy and international politics. The present danger to both endeavours is "post-modernism".

> The Western Rationalistic Tradition, the stipulation that theories be tested against empirical data, is no panacea for the prevention of evil. Nevertheless it does offer the best hope for academicians to make a positive contribution to the larger society because it can in some instances suggest a wise course for public policy and in others demonstrate that a policy is wrong. Post-modernism, in contrast, in its more extreme versions provides no such check. On the contrary, it leads directly to nihilism which can produce an intense and burning flame but which hardly moves society towards peace and justice (Krasner 1996: 125).

While Stephen Krasner's verdict may be extraordinarily harsh, the refusal of post-modernist, or, more general, post-positivist interventions is quite usual in IR. Thus the publishers of the influential journal *International Organization* explain that little of this debate was published in *IO*, since postmodernism denies the "use of evidence to adjudicate between truth claims". As Peter J. Katzenstein, Robert O. Keohane and Krasner (1998: 678) go on to declare: "postmodernism falls clearly outside of the social science enterprise, and in international relations research it risks becoming self-referential and disengaged from the world." Once again, the linkage between 'true' science and 'real world' relevance can hardly be dismissed.

The principles of proper research are laid down, prominently, by Gary King, Sidney Verba and Keohane in a much quoted textbook. Here, one can read that social science research occurs within a "well-established" and "stable structure of rules" (King et al. 1994: 12). By implication, the rules are neither in need of justification nor negotiable. Those who want to discuss them and their maybe unintended consequences must be content with the casual information that they will "have to look elsewhere for inspiration or for paradigmatic battles in which to engage" (7). Having in mind Wendt's warning of Balkanization, the latter is presumably not what scholars of IR are supposed to do. Rather they are urged to follow the rules which are valid for "all research where the goal is to learn facts about the real world" (6). The rules tell us, for instance that "all hypotheses need to be evaluated empirically before they can make a contribution to knowledge", that a research program "should deal with a significant real world topic", that a topic which "cannot be refined into a specific research project permitting valid descriptive or causal inference should be modified along the way or abandoned", and that theories must be "falsifiable" (16-19).

King et al. frequently refer to the real or external world. As we saw, Wendt put emphasis on an independent world as well, but takes trouble to re-build the distinction between theory and reality after accepting several specific features of social kinds. For King et al., there seems to be no problem at all. In their story, the "world that social scientists study is made up of particulars: individual voters, particular government agencies, specific cities, tribes, groups, states, provinces, and nations". The task of social science consists in the attempt "to go beyond these particulars to more general knowledge", and the "specific entities of the social world – or more precisely, special facts about these entities – provide the basis on which generalizations must rest" (35).

Katzenstein et al., while likewise paying little notice to the possibility that regularities may indicate some regular practices contaminated with power/rule, at least take up the argument that theories become part of the world. "Critical constructivists insist that scholars' work has normative consequences". What follows is a somehow queer counterexample. "A scientist may try to find a cure for cancer or instead develop a more virulent strain of anthrax. This choice, however, does not alter the mechanisms that cause cancer or anthrax". On the other hand, they append a sort of concession. "Critical constructivists, however, understand their project not simply as revealing relationships that exist independent of the investigator, but also as having the potential to alter these relationships themselves" (Katzenstein et al. 1998: 677).

Arguably, the crux of the matter is not the potential of critical constructivists to alter relationships, rather than the social quality of these relationships and the constitutive quality of intellectual projects or theories in general, no matter how they are labelled. Yet any such an approach seems to be prone to undermine the epistemic authority of IR and the capacity of the discipline to instruct policy-making. As Michael Nicholson remarks, it follows from the positivist's concern with generalisation "that the world is viewed as having some pattern to it" which "can be discerned if we try hard enough". And while he writes that this "does not logically imply that the behaviour of the world can be altered in foreseeable directions", he emphasizes that "it is hard to see how we can have any control" of the social world if it "does not have these characteristics". So conceived, it is not only the legacy of positivism, but also society's interest in problem-solving knowledge that presses scholars to make generalisations about international life: "policy requires positivism" (Nicholson 1996: 142). In the words of Ernst-Otto Czempiel, who stresses e.g. the "fact" – demonstrated by political science – that the "existence of democratic systems" is the "necessary and most important condition for the emergence of zones of peace" (1999: 86): "[wise power] must know what means and instruments in the interplay with the strategies of other agents produce the intended results" (155; my translation).

Facts and Norms

What happen to people and their relations in the stories we tell about world politics when we take it as a matter of course that there are general patterns and facts that we can discern? A research topic which has attracted much attention is the role of "norms" in international relations. Moreover, the study of norms has been used to question some epistemological and methodological assumptions of the discipline (Kratochwil/Ruggie 1986). But in the language of the mainstream, critics have "failed to produce the empirical research necessary to shake the rationalist and behavioral assumptions of neoliberal theory" (Katzenstein 1996: 21). In consequence, norms became "causal factors" or "independent variables" in order to show scientifically that they matter. Ronald L. Jepperson, Wendt, and Katzenstein (1996: 54) claim to employ the concept in a sociologically standard way. "Norms are collective expectations about proper behavior for a given identity". While, in my opinion, it seems doubtful that many sociologists would accept this statement, it has become a standard definition in IR.

Katzenstein exemplifies an explanation with "norms". What "force caused Governor Michael Dukakis, the Democratic candidate for president in 1988", he asks, "to dress up in military fatigues and ride around on a tank – looking foolish in the process – to demonstrate his toughness on the issue of national defense"? His answer: "collectively shared expectations of the American public about the military toughness of presidential candidates are what made the governor behave the way he did". And he concludes that "collective expectations can have strong causal effects" (Katzenstein 1996: 7).

I have no idea whether Dukakis succeeded in demonstrating a toughness on the issue of national defense, but I am pretty sure that Katzenstein fails to prove what made the governor behave they way he did. More precisely, Katzenstein's explanation is trivial, a simple conjecture or interpretation. The argument can be supported by a position which claims that rationalization is a species of ordinary causal explanation. "Whenever someone does something for a reason", Donald Davidson (1963: 685) writes, "he can be characterized as (a) having some sort of pro attitude toward actions of a certain kind, and (b) believing (or knowing, perceiving, noticing, remembering) that his action is of this kind". In most cases, we mention either a pro attitude or a belief when we explain our actions: "if you say you are biting your thumb at me because you want to insult me, there is no point in adding that you think that by biting your thumb at me you will insult me" (688). And as Davidson remarks, it is not only possible that a person has a reason for an action and doesn't perform the action, but also that he has a reason for an action, and performs the action, and yet this reason is not the reason why he did it (691).

Katzenstein assumes that Dukakis wants to demonstrate his toughness on the issue of national defense. Maybe, the governor desires to feel like a military hero and believes that riding on a tank is an action of that kind. In other words, 'one and the same' behavior can be described in different ways. If you tell me that the governor is riding around on a tank because he wants to demonstrate his toughness on the issue of national defense, I don't need to be told that you think that the governor believes that riding around on a tank will demonstrate his toughness. We can describe the behaviour of the candidate for president in terms of a teleological explanation or a strategic action: the presidential candidate wants to be elected president by the American public. Therefore, he tries to calculate the behaviour of the American electorate. On the one hand, he could try to discover a constant conjunction between the behaviour of candidates and the respective election results; even though it seems not very likely that he will find a corresponding regularity: 'presidential candidates (not) dressing up in military fatigues and riding around on a tank (do not) get elected'. On the other hand, he can draw on opinion polls or scientific measurements of public attitudes. Eventually, the candidate will learn that the majority of the electorate wants to have a president who is tough on the issue of military defence. Now he can think about strategies to make voters believe in his toughness.

But Katzenstein says that collectively shared expectations of the American public about the military toughness of presidential candidates made the governor behave they way he did. Obviously, something is missing. That is to say, we are left, e.g., without an account of the reason of the candidate for taking the course of action as well as an account of how rules work. Is there any norm according to which candidates for president should *dress up* in military fatigues? Does the American public collectively share the expectation that candidates for president must *demonstrate* their military toughness by riding around in a tank? Maybe, there is a norm to be found which reads that presidential candidates, presidents and other American politicians must *be* tough on the issue of national defence. When American politicians do not follow the rule, other agents (IR professors, journalists, and so on) may criticize them.

Nevertheless, Katzenstein's example is telling. He discriminates between norms having "'constitutive effects' that specify what actions will cause relevant others to recognize a particular identity" and norms having "'regulative effects' that specify standards of proper behaviour" (Katzenstein 1996: 5). While, for me, it is hard to make sense both of his formulation and distinction, I feel save to say that Katzenstein's example fits well to neither of his definitions. In other words, we will hardly find a norm which tells candidates for president that they should or must dress up in military fatigues and ride around on a tank in order to be recognized as (good, true?) presidential candidates or to behave properly. As Katzenstein's description already suggests, we can assume that the candidate wants to demonstrate his military toughness by doing what he did, like acting on a stage (but "looking foolish in the process"). On the one hand, then, it seems not very convincing to start with a world "made up of particulars" (King et al.), since the strategic action makes sense only in an intersubjective framework of social relations; e.g. of elections, voters, and candidates. On the other hand, the (assumed) course of action of the presidential candidate resembles the account of the social world in IR: the collective expectations of the American electorate were taken as given "thing" and not as something that is open to change.

Now, it has become routine in IR to say that agents make and change social facts. "Agents are not simply the puppets of social structure, since they can actively challenge the validity claims inherent in any communicative action." As Thomas Risse (1999: 10) goes on to write: "At the same time, they are social agents that produce and reproduce the intersubjective structures of meaning through their communicative practices". Henning Boekle, Volker Rittberger and Wolfgang Wagner (2001: 77) assume that a "socializee is quite able to reflect and even to modify in content the social norms and roles that he internalizes" (my translation).

At the same time, Boekle et al. write about the "explanatory power" of their "constructivist foreign policy theory", tested in terms of predictions of foreign policy behaviour. In their approach norms are an "independent variable", and they discuss processes of socialization as a "causal mechanism" that links social norms with foreign policy behaviour (73). How can a norm be an "independent variable" when agents (are able to) modify the content of the norm? Sebastian Harnisch (2003: 330; my translation) writes that "social agents and their actions are to be considered not only as products (in positivist terminology: 'dependent variable"), but also as constitutive for their coming into being ('independent variable')". One might think that "constitutive" brings into play the production of meaning by "signifying practices" (cf. Hall 2001). But in IR norms have a "regulative effect" on behaviour by increasing or decreasing the costs of certain ways of behaviour, and a "constitutive effect", by marking certain aims of action as "legitimate", that is, "norms have effects as motives" (Boekle et al. 2001: 75; my translation). In the words of Risse:

> Following Nicholas Onuf, Fritz Kratochwil, and Alexander Wendt, constructivism points to the constitutive role of ideational factors. Socially shared ideas – be it norms ('collective expectations about proper behavior for a given identity') or social knowledge about cause-and-effect relationships – not only regulate behavior but also constitute the identity of actors (Risse 2000: 5).[7]

Both regulation and constitution work in one direction, from "structures" or "socially shared ideas" to actors. What happens to agency? Boekle et al. inform us that they do not tackle the "reverse effect of German foreign policy 'agency' on the (normative) social structures of the international society or the German national society respectively", since they aim at "formulating predictions about

[7] This is an astonishing reading of Onuf and Kratochwil, since they link norms/rules to speech acts. Moreover, Onuf (1998) demonstrated the manifold defects of the definition of norms as "collective expectations", while Kratochwil refuted an understanding of norms as causes and questioned the self-evident connection of ideas and norms in IR (Kratochwil 2002: 46-49).

German foreign policy" (Boekle et al. 2001: 81; my translation). Obviously, the predictions depend on actors not modifying norms. What about situations that confront actors with conflicting norms or in which it is contested if a particular behaviour matches a certain standard? And why do we follow norms?

Boekle et al. (2001: 74) refers to a "logic of appropriateness", which says that "behaviors [...] are intentional but not wilful. They involve fulfilling the obligations of a role in a situation, and so of trying to determine the imperatives of holding a position". Risse and Kathryn Sikkink (1999: 8) give a similar explanation of the effects of norms: "people sometimes follow norms because they want others to think well of them, and they want to think well of themselves". Thus, one can read in a book on the "power of human rights": "at the end of a succeeding process of socialization the norms become second nature to the agents" (Risse et al. 2002: 21; my translation). "Actors follow the norm, because it is the normal thing to do" (Risse/Sikkink 1999: 17).

Why do we follow one norm and not another? Here, Frank Schimmelfennig points to two mechanisms of socialization. In the "mechanism of imitation" the "agents of socialization play role-models, which are observed and imitated by the socializee's, and the agents of socialization demonstrate behaviours, which are practised in role-playings". One might still ask why the mechanism does not work to the reverse. The second mechanism helps us out. In the "mechanism of persuasion", the "constitutive schemata and rules have the status of validity claims, which are argumentatively justified and disseminated in discourses of the agents of socialization with the socializee's". He concludes: "if the agents of socialization have the 'better arguments', the socializee's let themselves be convinced of the rightness of the validity claims and change their identity and interests accordingly: they internalize the schemata and rules" (Schimmelfennig 2003: 410; my translation). By implication, the things we do, normally, are the right ones, too. The "better argument" is the final cause.

Causes and Reasons

King, Keohane and Verba draw a distinction between causes and reasons. They (1994: 36-7) affirmatively cite John Ferejohn who argues that "we want social science theories to provide causal explanations of events and to give an account of the reasons for or meanings of social actions". That is to say that "we want to know not only what caused the agent to perform some act but also the agent's reason for taking the action". But if we should ask for the cause *and* for the reason of some action, we may come to the conclusion that President P declares war on State S because State S builds weapons of mass destruction *and* that President P declares war on State S for the reason of bringing democracy to State S. By implication, if the event that State S builds weapons of mass destruction in fact causes the event that President P declares war, then the reason to bring democracy to State S by declaring war on State S is of no importance in the physical world. Reasons become mere justifications or rationalizations, totally irrelevant for the chain of events.

Smith advances constitutive theorizing as an alternative to causal theorizing. He refers to Ludwig Wittgenstein and a distinction between reasons and causes. "The main difference is that, for Wittgenstein, to ask for a reason for action is to try and find the rule that led to the action, whereas to ask for a cause is to trace the mechanism, or find a statistical regularity, or a historical origin." However, the argument provided by Smith might be understood as leading to a conclusion that resembles the implication of the distinction between reasons and causes in King et al., since he refuses the view that reasons can be causes and writes that "finding a cause would say nothing about the reason" (Smith 2000: 159).

Wight defends the position that reasons are or should be understood as causes. In his opinion, this has come to be seen as necessary in the philosophy of social science in order to preserve the difference between action and behaviour. "For if the reason for an act is not part of the causal complex responsible for the act, then the contrast drawn between an act and a bodily movement, upon which hermeneutic accounts insists, is negated". For instance, the "difference between a waving arm and signalling to a friend depends upon the possession, by an agent, of a reason to wave one's arm in that manner, namely, the desire to signal to a friend". Hence, according to Wight, if "reasons are stripped of their causal function, behavioralism beckons" (Wight 2002: 38).

Onuf repudiates a simple correspondence between reasons and causes. But he does not say that reasons can not be causes, since he can identify innumerable occasions in which it makes sense for him to say that his reasons are causes for his conduct. "Unless I and others could assume as much, it is difficult for me to think why any of us should worry about reasons at all" (Onuf 2001: 250). Language enables us to think about actions two ways. "There are actions that we may or will perform, and those that we should or must perform." Language also enables us to jump back and forth across the gap. "A reason becomes a cause with a turn of phrase: 'I should write Karin because …' 'I will write her because …' " (252).

Reading Wittgenstein, I find little support for the claim that reasons can not be causes. Rather, in *The Blue Book*, he writes about the confusion between reason and cause; and that one "is led into this confusion by the ambiguous use of the word 'why' ".

> If, e.g., to the question, "why did you paint just this colour when I told you to paint a red patch?" you give the answer: "I have been shown a sample of this colour and the word 'red' was pronounced to me at the same time, and therefore this colour now always comes to my mind when I hear the word 'red' ", then you have given a cause for your action and not a reason (Wittgenstein 1958: 15).

Note that this description resembles the explanation with norms in IR, not least the "mechanism of imitation". As Wittgenstein writes, the "proposition that your action has such and such a cause, is a hypothesis". Such a hypothesis "is well-founded if one has had a number of experiences which, roughly speaking, agree in showing that your action is the regular sequel of certain conditions which we then call causes of the action". Yet, "in order to know the reason which you had for making a certain statement, for acting in a particular way, etc., no number of agreeing experiences is necessary, and the statement of your reason is not a hypothesis" (15). How does one give a reason for acting in a certain way?

> Suppose I pointed to a piece of paper and said, to someone: "this colour I call 'red'". Afterwards I give him the order: "now paint me a red patch". I then ask him: "why, in carrying out my order, did you paint just this colour?" His answer could then be: "This colour (pointing to the sample which I have given him) was called red; and the patch I have painted has, as you see, the colour of the sample". He has now given me a reason for carrying out the order in the way he did (14).

The difference between an explanation with causes and the giving of reasons comes up in the following remarks. "Giving a reason for something one did or said means showing a way which leads to this action. In some cases it means telling the way which one has gone oneself; in others it means describing a way which leads there" (14). Admittedly, there are accounts of causation that reject the Humean account of event-regularities (cf. Wight 2002). But Wittgenstein points out that the giving of reasons not only includes the description of ways which one in fact has gone: it includes statements of ways which one actually went as well as which one could have gone (Wittgenstein 1967: 22); and this is hardly something that we do call a causal explanation. Admittedly, there are occasions in which we want to know the reason for something someone did, and not a reason for acting in a certain way.[8] One might also argue that "to find the rule that led to the action" implicates a causal account of reasons, since there may be different ways which lead to the action. But, in my view, Wittgenstein is not interested in (causal) explanations of actions, rather than in opening up ways of action, rules and concepts to contestation and deliberation.[9]

As we saw, anyone who wants to get a hearing in IR is urged to subscribe to substantive theorizing about the real world. Those who do not develop such a research program "will remain on the margin of the field" (Keohane 1989: 173). We can take this remark as a statement of facts or a hypothesis respectively. I

[8] These occasions are stressed in criticism of (post-)Wittgensteinian approaches, cf. Douglas V. Porpora (1987: 64).

[9] On Wittgenstein's remarks on reasons and causes, cf. Arno Ros (1990). In his reading, a giving of reasons ("*begründende Erklärung*") should lead to the unforced establishment of a rule (334).

may say: 'I will develop such a research program because I don't want to remain on the margin of the field'. Alternatively, one might jump back across the gap: 'I should develop such a research program because ...'. Now, one might ask for as well as offer reasons for theorizing in a certain way. To say that 'we scholars of IR will remain on the margin of the field of IR if we do not follow certain rules of scientific research' is a strange argument – do not we scholars of IR constitute the margin as well as the core of the field? To say that '*those* who do not follow the norms of scientific research will remain on the margin of the field' is not so strange. If I do develop such a research program, others might explain my action by conjecturing that I do not want to remain on the margin of the field, that I believe that by developing the research program I will accomplish this goal and that this pair of want and believe is the cause of my action. Scholars of IR might say that collective expectations about a proper behaviour caused Jörg Meyer to develop such a research program. Intentionally, but not wilfully, he tries to hold (or to attain) a position. In the end, we may come to the result that we follow the norms of proper research because it is the normal thing to do.

I do not want to rebuke anyone for following the rules of scientific research in order to hold a position. Yet, on the one hand, if I place myself in the "middle ground", then I strengthen the rules of scientific research in IR, and the margin of the field becomes a little bit more marginal (quantitatively). I do not have to assume that the "geography" of IR confronts me as an objective social fact – I just don't care about it as well as about the concept of science and the rules for the production of knowledge, but only about my position. Obviously, there is no guarantee that I will be successful. On the other hand, if I do indeed theorize in this way, that is to say, if I make generalisations, give causal explanations and provide predictions, I leave out the capacity of people of saying no, of beginning something anew and of talking together about their relations and transforming them, both in my stories about IR and about international relations.

Works Cited

Arendt, Hannah (1958). *The Human Condition*. Chicago: University of Chicago Press.

Arendt, Hannah (1969). *On Violence*. New York: Harcourt, Brace and World.

Boehkle, Henning, Volker Rittberger and Wolfgang Wagner (2001). "Soziale Normen und normgerechte Außenpolitik. Konstruktivistische Außenpolitiktheorie und deutsche Außenpolitik nach der Vereinigung". *Zeitschrift für Politikwissenschaft* 1, 71-103.

Czempiel, Ernst-Otto (1999). *Kluge Macht: Außenpolitik für das 21. Jahrhundert*. München: Beck.

Davidson, Donald (1963). "Actions, Reasons, and Causes". *The Journal of Philosophy* 60.23, 685-700.

Debrix, Francois, ed. (2003). *Language, Agency, and Politics in a Constructed World*. Armonk & London: M. E. Sharpe, 3-25.

Derrida, Jacques (1992). "Force of Law: The 'Mystical Foundation of Authority'". *Deconstruction and the Possibility of Justice*. Ed. D. Cornell, M. Rosenfeld & D. Carlson. New York: Routledge.

Doty, Roxanne Lynn (2000). "Desire All the Way Down". *Review of International Studies* 26, 137-39.

Hall, Stuart, ed. (2001). *Representation. Cultural Representations and Signifying Practices*. London: Sage.

Harnisch, Sebastian (2003). "Theorieorientierte Außenpolitikforschung in einer Ära des Wandels". *Die neuen Internationalen Beziehungen*. Ed. G. Hellmann, K. D. Wolf & M. Zürn. Baden-Baden: Nomos, 313-60.

Jepperson, Ronald L., Alexander Wendt & Peter J. Katzenstein (1996). "Norms, Identity, and Culture in National Security". *The Culture of National Security: Norms and Identity in World Politics*. Ed. P. J. Katzenstein. New York: Columibia UP, 33-75.

Katzenstein, Peter J. (1996). "Introduction." *The Culture of National Security: Norms and Identity in World Politics*. Ed. P. J. Katzenstein. New York: Columibia UP, 1-32.

Katzenstein, Peter, Robert Keohane & Stephen Krasner (1998): "*International Organization* and the Study of World Politics". *International Organization* 52.4, 645-85.

Keohane, Robert O. (1989). "International Institutions: Two Approaches". *International Institutions and State Power*. Ed. R.O. Keohane. Boulder: Westview, 158-79.

King, Gary, Robert Keohane & Sidney Verba (1994). *Designing Social Inquiry. Scientific Inference in Qualitative Research*. Princeton: Princeton UP.

Krasner, Stephen (1996). "The Accomplishments of International Political Economy". *International Theory: Positivism and Beyond*. Ed. S. Smith, K. Booth & M. Zalewski. Cambridge: Cambridge UP, 108-127.

Kratochwil, Friedrich V. (2002). "Souveränität und Moderne. Eine begriffliche Analyse des semantischen Feldes". *Regieren in internationalen Institutionen*. Ed. M. Jachtenfuchs & M. Knodt. Opladen: Leske & Budrich, 29-51.

Kratochwil, Friedrich V. & John G. Ruggie (1986). "International Organization: A State of the Art on the Art of the State". *International Organization* 40.4, 753-775.

Mayer, Peter (2003). "Die Epistemologie der Internationalen Beziehungen: Anmerkungen zum Stand der 'Dritten Debatte". *Die neuen Internationalen Beziehungen*. Ed. G. Hellmann, K. D. Wolf & M. Zürn. Baden-Baden: Nomos, 47-97.

Nicholson, Michael (1996). "The Continued Significance of Positivism?" *International Theory: Positivism and Beyond*. Ed. S. Smith, K. Booth & M. Zalewski. Cambridge: Cambridge UP, 128-45.

Onuf, Nicholas G. (1998). "Everyday Ethics in International Relations". *Millenium: Journal of International Studies* 3, 669-93.

Onuf, Nicholas G. (2001). "*The Politics of Constructivism*". *Constructing International Relations: the Next Generation*. Ed. K. M. Fierke & K. E. Jörgensen. Armonk & London: M.E. Sharpe, 236-254.

Porpora, Douglas V. (1987). *The Concept of Social Structure*. New York: Greenwood Press.

Risse, Thomas (1999). "'Let's Argue!': Communicative Action in World Politics". *International Organization*, 54.1, 1-39.

Risse, Thomas, Anja Jetschke & Hans Peter Schmitz (2002). *Die Macht der Menschenrechte. Internationale Normen, kommunikatives Handeln und politischer Wandel in Ländern des Südens*. Baden-Baden: Nomos.

Risse, Thomas & Sikkink, Kathryn (1999). "The Socialization of International Human Right Norms into Domestic Practices: Introduction". *The Power of Human Rights. International Norms and Domestic Change*. Ed. T. Risse, S. Ropp & K. Sikkink. Cambridge: Cambridge UP, 1-38.

Ros, Arno (1990). *Begründung und Begriff. Band 3: Moderne*. Hamburg: Meiner

Schimmelfennig, Frank (2003). "Internationale Sozialisation: Von einem 'erschöpften' Forschungsprogramm zu einem produktiven Forschungsprogramm?" *Die neuen Internationalen Beziehungen*. Ed. G. Hellmann, K. D. Wolf & M. Zürn. Baden-Baden: Nomos, 401-27.

Smith, Steve (1995). "The Self Images of a Discipline: A Genealogy of International Relations Theory". *International Relations Theory Today*. Ed. K. Booth & S. Smith. Cambridge: Polity, 1-37.

Smith, Steve (2000). "Wendt's World". *Review of International Studies* 26, 151-163.

Tully, James (2002). "Political Philosophy as a Critical Activity". *Political Theory* 30.4, 533-55.

Waldenfels, Bernhard (1998). "Comment on John Searle's *The Construction of Social Reality*". *Analyse & Kritik* 20, 159-65.

Walker, R.B.J. (1993). *Inside/Outside: International Relations as Political Theory*. Boulder: Lynne Rienner.

Wendt, Alexander (1999). *Social Theory of International Politics*. Cambridge: Cambridge UP.

Wendt, Alexander (2000). "On the Via Media: a Response to the Critics". *Review of International Studies* 26, 165-80.

Wittgenstein, Ludwig (1958). *The Blue and the Brown Books*. Oxford: Basil Blackwell.

Wittgenstein, Ludwig (1967). *Lectures and Conversations on Aesthetics, Psychology and Religious Belief*. Ed. Cyril Barrett. Berkely: University of California Press.

Wight, Colin (2002). "Philosophy of Social Science and International Relations". *Handbook of International Relations*. Ed. W. Calsnaess, T. Risse & B. A. Simmons. London: Sage, 23-51.

Contributors

Steven L. Arxer was born in Miami, Florida (USA) in 1980. He received his M.A. degree from the University of Miami. Currently, he is a Ph.D. student in Sociology at the University of Florida. His interests are sociological theory, Latin American studies, and globalization.

Anke Bartels read English and German Studies at the Technical University of Braunschweig. After receiving her PhD, she was a visiting lecturer for German Studies at Strathclyde University in Glasgow. She is currently working on a post-doc thesis on East African women writers as well as teaching English literature and cultural studies at the University of Magdeburg. Her research interests include contemporary British theatre, feminism, cultural studies and post-colonial literatures.

Linda L. Belgrave was born in Cleveland, Ohio (USA) in 1950. She received her Ph.D degree from Case Western Reserve University in Sociology. Currently, she is Associate Professor of Sociology at the University of Miami. Her areas of interest are social gerontology, medical sociology, and research methods.

Orly Benjamin is a senior lecturer at the department of sociology and anthropology in Bar-Ilan university, Israel. Her fields of research are emotions, gender, couple relationships, temporary employment, job insecurity and the body. In her various fields of research she aims at a theory of silencing as a characteristic of social and political discourses and its challenge by active unsilencing.

Steven Corcoran is a translator and doctoral candidate at the University of New South Wales, Australia. His PhD examines the political thought of Michel Foucault, Jacques Rancière, and Alain Badiou.

Catarina Frois is a doctoral student in Anthropology at the Institute of Social Sciences, Lisbon, Portugal. She is a Foundation for Science and Technology Scholar, Member of the European Association of Social Anthropologists and Member of the Portuguese Association of Anthropologists. Catarina Frois has published in several referred journals and participated in national and international multidisciplinary summits. Her main interests for research are: cultural aspects of anonymity, privacy, surveillance and identification.

AGNIESZKA KOŁODZIEJ-DURNAŚ, (Ph.D., Institute of Sociology and Psychology, University of Szczecin, Poland), born 1973, studied sociology and English. Main research interests: contemporary sociological theories, sociology of organization, sociology of knowledge, cultural differences. Teaches in Polish and English. The author of a book on A.Giddens' structuration theory, co-author and editor of a book on the corporate culture of selected companies and institutions.

TARJA-LIISA LAAKSONEN, born in 1957, is a PhD student (Organizational Sociology) at the Helsinki School of Economics, Finland. She obtained her M.Sc. (Economics) from the Helsinki School of Economics in 1986 and her M.Phil. (Development Studies) from the Institute of Development Studies, Sussex University, Brighton, U.K. in 1988. For 15 years she has worked in development cooperation with the Finnish bilateral cooperation, the United Nations, and the European Commission cooperation.

MARIA ANTONIETTA MARIANI, born in 1968, is an Anthropologist and Economist (Degree in Economics; Master in Communication; Master in Ethno-Anthropological Science). She is a researcher for the Censis Foundation on the following issues: local development, urban transformation, migration, industrial districts etc. At the moment, she is working on Space and Representation: sharing and engraving spaces; conflicts and borders; city and space; migrating people, nomads' and refugees' space.

JÖRG MEYER, born 1965; Ph.D.; read Political Science, Economics and Modern History at the Technical University of Braunschweig. Currently, he teaches International Relations and Political Theory and is co-responsible for the Master Programme in "Peace and Conflict Studies" at the University of Magdeburg. His research interests include International Relations Theory, Political Theory and Language and contested concepts in contemporary world politics.

JOHN W. MURPHY was born in Youngstown, Ohio (USA) in 1948. He received his Ph.D. degree in Sociology from Ohio State University. Currently, he is Professor of Sociology at the University of Miami. His interests are social philosophy, sociological theory, and globalization.

HOGNE ØIAN, born in 1960, is associate professor at the Department of Social Anthropology, University of Tromsö, Norway. Latest publication: "Time out and Drop out. On the relation between linear time and individualism". *Time and Society* 13.2/3, 2004. Research interests: Linear time, forms of individualism, consumption, family- and kinship-relations, gender.

Rocío Orsi Portalo, born in 1976, is Assistant Teacher at the Universidad Carlos III de Madrid. He graduated in Philosophy at the Universidad Autónoma Madrid (1998) and completed his Ph.D. thesis on *Thought and Action in Sophocles* in October, 2005. His research fields include Moral and Political Philosophy and History of Philosophy.

Leena Petersen, born in 1977, studied Aesthetics/Cultural Studies, Philosophy, Literature and Scandinavian Studies at the Humboldt University in Berlin and completed her MA in Social and Political Thought at the University of Sussex (supported by the Heinrich Böll Foundation). Currently she is working on her PhD thesis in Philosophy at the University of Sussex regarding the form of imagistic approaches in the Fin de Siècle.

Dagmar Reichardt was born in Rome/Italy in 1961. Assistant Professor for Italian and Comparative Literary and Cultural Studies at the University of Bremen, Germany. She studied Romance languages and literature, philosophy, modern German literature and art history in New York, Urbino, Frankfurt a.M. and Hamburg where she began her teaching career and founded the arts journal *Zigzag Italia*. Principal working fields: Literary Modernism in Italy (16th, 19th-21st centuries), Cultural and Postcolonial Studies, comparatistics, narratology, imageology, discourse analyses and production aesthetics. Since 2001 panelist of Premi Internazionali Flaiano, since 2006 member of the academic advisory board of A.I.P.I. (Associazione Internazionale Professori di Italiano). Currently serving on a postdoctoral thesis on discourses of power in francophone theatre.

Holger Rossow (*1970) works as a Senior Lecturer of Cultural Studies at the University of Rostock, Germany. His current research interests are globalisation, the discourse of globalisation and transatlantic relations. Among his latest publications are articles on the issue of regional and national cultures in the British Isles, the development of transatlantic relations in the context of globalisation and the issue of a new international order.

Sarit Sambol completed her MA degree at the gender studies program in Bar-Ilan university. Her thesis was on "Processes of change among Israeli women from a lower economic background who participate in an economic empowerment workshop". She is the co-author of several articles and will soon start her Ph.D. Studies in Canada.

Ulf Schulenberg (born 1966) studied American Studies, English and French Literature at the University of Bremen and at U.C.L.A. (USA). He received his doctorate (American Literature) from the University of Bremen in 1999, and in 2005 he finished his postdoctoral thesis ("Habilitation"). He has been a Visiting Scholar at Cornell University and at New School University. His main fields of interest are cultural and literary theory, aesthetic theory, the American cultural left, and the modern and postmodern American novel.

Michael Schultze, born in 1973, studied Political Science, English and History. Currently he is working as a lecturer at the Department of Political Science at the University of Magdeburg. His research interests include International Relations Theory, Peace and Conflict Studies and Sub-Saharan Africa.

Agata Stopińska, born in 1975, studied applied linguistics at Poznań University (Poland), as well as German and English studies in Freiburg im Breisgau (Germany). Since 2001, she has been a lecturer and coordinator of "European Studies" at the University of Magdeburg. Currently she is working on her PhD project on gender representations in the Irish daily press and discourse studies.

Jatin Waglé has been teaching English in Mumbai, India. Currently, he is pursuing doctoral research on the reception of the purportedly untranslatable writings of Theodor Adorno, at the University of Hannover. He is also a fellow of the Heinrich Böll Stiftung, Germany.

Dirk Wiemann teaches English literature and cultural studies at the University of Magdeburg. He has published on Indian writing in English, writers in exile in the Britain of the 1930s, on crime fiction, pop music and the culture industry during the war on Iraq. Recently, he has finished his post-doctoral thesis on the transculturation of nation and temporalities in postcolonial Indian writing.

Transpekte: Transdisziplinäre Perspektiven der Sozial- und Kulturwissenschaften / Transpects: Transdisciplinary Perspectives of the Social Sciences and Humanities

Herausgegeben von Johannes Angermüller, Anke Bartels, Dietmar Fricke, Raj Kollmorgen, Jörg Meyer und Dirk Wiemann

Band 1 Dirk Wiemann / Agata Stopińska / Anke Bartels / Johannes Angermüller (eds.): Discourses of Violence – Violence of Discourses. Critical Interventions, Transgressive Readings, and Post-national Negotiations. 2005.

Band 2 Michael Schultze / Jörg Meyer / Britta Krause / Dietmar Fricke (Hrsg.): Diskurse der Gewalt – Gewalt der Diskurse. 2005.

Band 3 Anke Bartels / Michael Schultze / Agata Stopińska (eds.): Re/defining the Matrix. Reflections on Time – Space – Agency. 2006.

Band 4 Britta Krause / Tania Meyer / Nina Pippart / Dietmar Fricke (Hrsg.): Chronotopographien. Agency in ZeitRäumen. 2006.

www.peterlang.de